NUCLEAR MATERIALS AND DISASTER RESEARCH

ADVANCED REACTORS

REVIEW OF U.S. EFFORTS IN THE DEVELOPMENT OF NUCLEAR AND FUSION ENERGY

Nuclear Materials and Disaster Research

Additional books in this series can be found on Nova's website under the Series tab.

Additional e-books in this series can be found on Nova's website under the e-book tab.

NUCLEAR MATERIALS AND DISASTER RESEARCH

ADVANCED REACTORS

REVIEW OF U.S. EFFORTS IN THE DEVELOPMENT OF NUCLEAR AND FUSION ENERGY

EMELIA CLARKE
EDITOR

nova publishers
New York

Library of Congress Cataloging-in-Publication Data

ISBN: 978-1-63463-263-8

Published by Nova Science Publishers, Inc. † New York

CONTENTS

PREFACE

The Department of Energy's (DOE) Office of Nuclear Energy's (NE) approach to advanced reactor research and development (R&D) focuses on three reactor technologies—high-temperature gas-cooled reactors, sodium-cooled fast reactors, and fluoride-salt-cooled high-temperature reactors—but NE is also funding research into other advanced reactor technologies. NE's approach is to conduct research in support of multiple advanced reactor technologies, while collaborating with industry and academia, with the ultimate goal for industry to take the results of NE's research to the next step of development and commercialization. This book describes NE's approach to advanced nuclear reactor R&D and examines how NE plans and prioritizes its advanced reactor R&D activities, including deploying an advanced reactor. This book also discusses the DOE's International Thermonuclear Experimental Reactor (ITER) Agreement. It examines how and why the estimated costs and schedule of the U.S. ITER Project have changed since 2006; the reliability of DOE's current cost and schedule estimates; and actions DOE has taken to reduce U.S. ITER Project costs and plan for their impact on the overall U.S. fusion program.

Chapter 1 – NE conducts R&D on advanced nuclear reactor technologies with multiple aims, including (1) improving the economic competitiveness of nuclear technology to ensure that nuclear power continues to play a role in meeting our nation's energy needs; (2) increasing safety; (3) minimizing the risk of nuclear proliferation and terrorism; and (4) addressing environmental challenges, such as reducing greenhouse gas emissions. External groups have been critical of NE for, among other things, how it prioritizes advanced reactor R&D. GAO was asked to review NE's advanced reactor R&D efforts. This report (1) describes NE's approach to advanced nuclear reactor R&D and (2) examines how NE plans and prioritizes its advanced reactor R&D activities, including deploying an advanced reactor. GAO reviewed laws and reports concerning NE's efforts to develop advanced reactor technologies and interviewed NE officials and a nonprobability sample of companies developing such technology, selected because of their involvement with DOE's R&D efforts.

Chapter 2 – ITER is an international research facility being built in France to demonstrate the feasibility of fusion energy. Fusion occurs when the nuclei of two light atoms collide and fuse together at high temperatures, which results in the release of large amounts of energy. The United States has committed to providing about 9 percent of ITER's construction costs through contributions of hardware, personnel, and cash, and DOE is responsible for managing those contributions, as well as the overall U.S. fusion program. In fiscal year 2014, the U.S. ITER Project received $199.5 million, or about 40 percent of the overall U.S. fusion program budget. GAO was asked to review DOE's cost and schedule estimates for the U.S. ITER Project. This report examines (1) how and why the estimated costs and schedule of the U.S. ITER Project have changed since 2006, (2) the reliability of DOE's current cost and schedule estimates, and (3) actions DOE has taken to reduce U.S. ITER Project costs and plan for their impact on the overall U.S. fusion program. GAO reviewed documents; assessed DOE's current estimates against best practices; and obtained the perspectives of 10 experts in fusion energy and project management.

In: Advanced Reactors
Editor: Emelia Clarke
ISBN: 978-1-63463-263-8

Chapter 1

ADVANCED REACTOR RESEARCH: DOE SUPPORTS MULTIPLE TECHNOLOGIES, BUT ACTIONS NEEDED TO ENSURE A PROTOTYPE IS BUILT*

United States Government Accountability Office

WHY GAO DID THIS STUDY

NE conducts R&D on advanced nuclear reactor technologies with multiple aims, including (1) improving the economic competitiveness of nuclear technology to ensure that nuclear power continues to play a role in meeting our nation's energy needs; (2) increasing safety; (3) minimizing the risk of nuclear proliferation and terrorism; and (4) addressing environmental challenges, such as reducing greenhouse gas emissions. External groups have been critical of NE for, among other things, how it prioritizes advanced reactor R&D.

GAO was asked to review NE's advanced reactor R&D efforts. This report (1) describes NE's approach to advanced nuclear reactor R&D and (2) examines how NE plans and prioritizes its advanced reactor R&D activities, including deploying an advanced reactor. GAO reviewed laws and reports concerning NE's efforts to develop advanced reactor technologies and

* This is an edited, reformatted and augmented version of the United States Government Accountability Office publication, GAO-14-545, dated June 2014.

interviewed NE officials and a nonprobability sample of companies developing such technology, selected because of their involvement with DOE's R&D efforts.

WHAT GAO RECOMMENDS

To better prepare DOE to meet the requirement of EPAct 2005 to deploy the NGNP prototype reactor, GAO recommends that DOE develop a strategy for resuming the NGNP Project and provide a report to Congress updating the status of the project. DOE agreed in principle with GAO's first recommendation and respectfully disagreed with the second. GAO believes these recommendations remain valid as discussed in the report.

WHAT GAO FOUND

The Department of Energy's (DOE) Office of Nuclear Energy's (NE) approach to advanced reactor research and development (R&D) focuses on three reactor technologies—high-temperature gas-cooled reactors, sodium-cooled fast reactors, and fluoride-salt-cooled high-temperature reactors—but NE is also funding research into other advanced reactor technologies. NE's approach is to conduct research in support of multiple advanced reactor technologies, while collaborating with industry and academia, with the ultimate goal for industry to take the results of NE's research to the next step of development and commercialization. This approach provides several advantages, including flexibility in responding to changes in future U.S. energy policy. Many representatives that GAO talked to from the nuclear power industry and the National Academy of Sciences agree with NE's approach, saying that current policies on controlling greenhouse gas emissions and disposing of nuclear waste do not make a compelling case for choosing a reactor technology to develop. However, others GAO talked to are critical of some of the reactor technologies NE chooses to research, citing economic and technological challenges. The Nuclear Energy Advisory Committee has criticized NE's approach, recommending that NE focus its efforts on a smaller number of technologies to help ensure that a reactor prototype is deployed. To remain aware of industry's R&D needs and international nuclear energy developments, NE regularly collaborates with industry and international organizations.

NE uses internal and external reviews to set program and funding priorities for advanced reactor R&D activities and to evaluate progress toward program goals. For example, NE conducts internal monthly and quarterly reviews to discuss project status, budgets, and technical highlights. Furthermore, NE's R&D efforts are periodically reviewed by external entities, including the Nuclear Energy Advisory Committee. Among the advanced reactor technologies that NE's R&D currently supports, the high-temperature gas-cooled reactor is the technology that is most likely to be deployed and commercialized in the near term, according to an NE planning document. NE officials said this likelihood is based on the wide range of potential industry market applications and because of substantial government investments in the technology's development. NE has been pursuing this technology under the Next Generation Nuclear Plant (NGNP) Project, as established by the Energy Policy Act of 2005 (EPAct 2005). Under EPAct 2005, DOE is to deploy a prototype reactor for NGNP by the end of fiscal year 2021. However, in 2011, DOE decided not to proceed with the deployment phase of this project, citing several barriers. For example, NE and industry have been unable to reach an agreement on a cost-share arrangement to fund the deployment phase because of a disagreement on the applicable cost-share levels and how and when the cost-share would be applied to specific activities or project phases. Although NE continues to conduct R&D for the NGNP Project, it has not developed a strategy to overcome the cost-share issue and other barriers to resuming the deployment phase of the project. Furthermore, DOE has not selected initial reactor design parameters or reported to Congress on an alternative date for making this selection. Without doing so, it is not clear when NE is going to take this next step in deploying the NGNP prototype reactor and it risks the project not being completed by the targeted date in 2021.

ABBREVIATIONS

DOE	Department of Energy
EPAct 2005	Energy Policy Act of 2005
FACA	Federal Advisory Committee Act
NE	Office of Nuclear Energy
NGNP	Next Generation Nuclear Plant
NRC	Nuclear Regulatory Commission

OECD	Organisation for Economic Co-operation and Development
R&D	research and development

***

June 23, 2014

The Honorable Dianne Feinstein
Chairman
Subcommittee on Energy and Water Development
Committee on Appropriations
United States Senate

Dear Madam Chairman:

Energy demand in the United States is expected to continue to grow over the coming decades, according to the U.S. Energy Information Administration, and energy security and greenhouse gas emissions from the burning of fossil fuels will continue to be of serious concern. Nuclear energy accounts for about 20 percent of electricity generation in the United States and produces no air pollution or greenhouse gases. However, the United States faces several challenges in developing additional nuclear energy facilities. For example, the March 2011 accident at Japan's Fukushima Daiichi commercial nuclear power plant has increased concerns about the safe operation of nuclear power plants worldwide. In addition, many uncertainties surround long-term storage and disposal options for spent nuclear fuel from existing power plants.[1] Spent nuclear fuel is considered one of the most hazardous substances on earth, and exposure can cause environmental harm and long-term health hazards, such as cancer, in those who are exposed to it. Spent nuclear fuel also raises concerns over nuclear proliferation and terrorism[2] Specifically, concerns exist that terrorist networks could divert nuclear material intended for peaceful purposes to the development of nuclear weapons. In addition, the cost of developing, deploying, and operating new nuclear power facilities in the United States, relative to other types of energy sources, has been an impediment to increasing nuclear production.

Research on a new generation of advanced nuclear reactors, which are intended to improve on the current generation of nuclear reactors through increased safety, improved economic competitiveness, and reduced nuclear waste, is being conducted by governments and private industry around the world. In the United States, the Department of Energy (DOE) spends more than $80 million per year to develop advanced nuclear reactors with the aim of eventually replacing aging U.S. commercial light water reactors,[3] many of which are operating past the end of their initial 40-year operating period under renewed operating licenses.[4] DOE's Office of Nuclear Energy (NE) conducts research and development (R&D) on advanced reactors with multiple aims, including (1) improving the economic competitiveness of nuclear energy to ensure that nuclear energy continues to play a role in meeting our nation's energy needs, (2) increasing safety, (3) minimizing the risks of nuclear proliferation and terrorism, and (4) addressing environmental challenges such as reducing greenhouse gas emissions.[5]

In September 2006, we reported on NE's progress in deploying an advanced reactor prototype,[6] stating that initial R&D results were favorable, but DOE officials considered the schedule to be challenging, given the amount of R&D that remained to be conducted. External groups, such as the Nuclear Energy Advisory Committee,[7] the National Academy of Sciences, and we have reported on NE's advanced nuclear reactor R&D efforts from 2008 to 2013.[8] These reports have been critical of how NE prioritizes its advanced reactor R&D, the advanced reactor technologies that NE has chosen to fund, and how NE partners with industry to advance program goals

In this context, you asked us to review NE's advanced reactor R&D efforts. Our objectives were to (1) describe NE's approach to advanced reactor R&D and (2) examine how NE plans and prioritizes its advanced reactor R&D activities, including deploying an advanced reactor prototype.

To address these objectives, we reviewed relevant federal laws and regulations, as well as reports by DOE, us, and others that describe NE's efforts to develop advanced reactor technologies. In addition, to better understand NE's approach to and decision-making processes for advanced reactor R&D, we interviewed NE officials; representatives from a nonprobability sample of companies developing advanced reactor technology;[9] knowledgeable members of the Nuclear Energy Advisory Committee and the National Academy of Sciences' National Research Council; and representatives from industry groups, including the Next Generation Nuclear Plant Industry Alliance and the Nuclear Energy Institute.[10] In addition, we visited two of DOE's national laboratories—Oak Ridge National Laboratory and Idaho National Laboratory—to better understand NE's R&D activities for advanced nuclear reactors. To select these laboratories, we analyzed the appropriated funding levels associated with NE across the 10 national laboratories that NE funds.[11] We selected the 2 laboratories with the highest levels of funding dedicated to NE programs over the past 5 years.

We conducted this performance audit from August 2013 to June 2014 in accordance with generally accepted government auditing standards. Those standards require that we plan and perform the audit to obtain sufficient, appropriate evidence to provide a reasonable basis for our findings and conclusions based on our audit objectives. We believe that the evidence obtained provides a reasonable basis for our findings and conclusions based on our audit objectives.

Background

According to its 2010 *Nuclear Energy Research and Development Roadmap: A Report to Congress*, NE's primary mission is to advance nuclear power as a resource capable of meeting the nation's energy supply, environmental, and energy security needs by resolving technical, cost, safety, proliferation resistance, and security barriers through research, development, and demonstration, as appropriate. NE conducts research aimed at (1) improving the reliability, sustaining the safety, and extending the operational lifetime of existing light water nuclear reactors; (2) supporting the development of the next generation of nuclear reactors, including light-water-reactor-based small modular reactors and advanced reactors, with a focus on affordability; (3) developing sustainable nuclear fuel cycles; and (4) reducing the risk of nuclear proliferation and terrorism. Light-water-reactor-based small modular reactors are smaller in size and energy output than conventional light water reactors—but use the same basic technology for the nuclear reactor core—and offer several potential advantages over existing light water reactors, including lower capital and construction costs through factory fabrication; enhanced safety and security; improved operation times and longer life cycles; and flexibility to be sited at locations that cannot support large nuclear plants, such as isolated areas or sites with limited water supplies.

Advanced reactors, including advanced small modular reactors, use innovative nuclear fuels, coolants, and energy systems, and offer the potential for significant advantages over existing light water reactors, including greater energy conversion efficiency, reduced plant size, lower construction and operation costs, and improved safety. NE has mainly conducted advanced reactor R&D on high-temperature reactors and fast reactors. High-temperature reactors produce electricity, as well as process heat that can be used for industrial purposes, such as refining petroleum or producing hydrogen, and replace current sources of process heat from burning natural gas or other fossil fuels, which emit greenhouse gases. Fast reactors can use spent nuclear fuel as their fuel source, which reduces the need for long-term storage of spent nuclear fuel, and would more efficiently use uranium, helping reduce nuclear waste.

NE conducts nuclear reactor R&D through its Reactor Concepts Research, Development, and Demonstration program, which aims to (1) help advance nuclear power as a resource capable of meeting the nation's energy, environmental, and national security needs and (2) develop new and advanced reactor designs and technologies that advance the state of reactor technology and improve the economic competitiveness of nuclear power. The program encompasses the following subprograms:

- The Light Water Reactor Sustainability subprogram is developing the scientific basis to extend existing nuclear power plant operating life beyond the current licensing period and ensure their long-term reliability, productivity, safety, and security. This subprogram conducts research into materials aging and degradation, updating instrumentation and controls, and assessing reactor safety margins, among other things.
- The Advanced Reactor Concepts subprogram[12] supports the development of innovative reactor technologies that may offer improved safety, functionality and affordability; more efficient energy conversion; increased proliferation resistance and security; and that build upon existing nuclear technology and operating experience. It supports research to reduce technical barriers for advanced nuclear energy systems, and its efforts support various reactor technologies at different maturity levels.
- The Advanced Small Modular Reactor subprogram[13] supports the development of innovative small modular reactor designs that potentially offer improved safety, functionality, and affordability; more efficient energy conversion; increased proliferation resistance and security; and simplified operation and maintenance. For example, the program supports research into novel sensors and control systems for multiple reactor units, as well as advanced materials development.

Other NE programs conduct R&D that supports the nuclear reactor R&D efforts. The Fuel Cycle R&D program—which is primarily responsible for developing sustainable nuclear fuel cycles,[14] innovative processes to recover uranium from spent nuclear fuel, and nuclear fuel storage options—also supports existing reactors by developing accident-tolerant fuel and advanced reactors by developing advanced proliferation-resistant fuels.[15] The Nuclear Energy Enabling Technologies program conducts R&D on technologies that directly support and complement NE's advanced reactor and fuel cycle R&D efforts. This program coordinates R&D on common issues and challenges that confront other NE R&D efforts to avoid duplication of effort.[16]

NE conducts most of this research at 10 DOE national laboratories across the country and, over the past 3 years, has spent an average of about $840 million per year on its mission to advance nuclear power (see table 1).[17] Of this amount, NE provides about $50 million annually to engage U.S. universities through its Nuclear Energy University Program to fund R&D and to build infrastructure and capabilities to enhance universities' ability to perform research and educate students.[18]

Table 1. Total Funding for DOE Office of Nuclear Energy (NE), Fiscal Years 2012 through 2014

Dollars in millions			
Program	**Fiscal year 2012**	**Fiscal year 2013**	**Fiscal year 2014**
Fuel Cycle Research and Development	$181.0	$169.9	$186.2
Nuclear Energy Enabling Technologies	71.3	67.9	71.1
Reactor Concepts Research, Development, and Demonstration	110.7	104.8	112.8
Small Modular Reactor Licensing Technical Support	67.0	62.7	110.0
Other NE programs, transfer from Department of State, and use of prior year balances[a]	423. 9[b]	393.0[b]	397.4
Total	**$853.8**	**$798.3**	**$877.5**

Source: DOE. | GAO-14-545

Note: Numbers may not add to total due to rounding.

[a] Other NE programs include facilities management, international cooperation, the Idaho Sitewide Safeguards and Security program, the Integrated University Program, and overall direction and execution of the Office of Nuclear Energy programs. The Integrated University Program funding is intended to support nuclear science and engineering by funding scholarships and fellowships in nuclear energy related fields of study, among other things. These funding levels

also include funds transferred from the Department of State, funds transferred to the Small Business Innovation Research and Small Business Technology Transfer programs outside of NE, and the use of NE's prior year balance.

[b] The Idaho Sitewide Safeguards and Security program was funded within Other Defense Activities for fiscal years 2012 and 2013, but are reflected this table. The program was funded within NE for fiscal year 2014. Other Defense Activities funds elements that relate to and support the defense oriented activities within DOE.

NE's recent advanced reactor R&D efforts began in 2000, when NE convened a group of senior governmental officials from nine countries to discuss the development of such reactors in the United States and internationally.[19] This group, called the Generation-IV International Forum, and the Nuclear Energy Advisory Committee produced a Technology Roadmap for Generation-IV Nuclear Energy Systems in 2002. The intent of the forum was to develop competitively priced and reliable nuclear reactors, while satisfactorily addressing nuclear safety, waste, and proliferation concerns. In response to these efforts, the United States determined that it would fund the development of a high-temperature gas-cooled reactor as its top priority.[20] In addition, NE funded research into sodium-cooled fast reactors—fast reactors that use sodium to cool the reactor core—as well as a variety of other advanced reactors.[21]

The Energy Policy Act of 2005 (EPAct 2005) established in law the Next Generation Nuclear Plant (NGNP) Project.[22] The purpose of the NGNP is to develop a prototype reactor using advanced technology to generate electricity, hydrogen, or both. The law states that the NGNP Project shall consist of research, development, design, construction, and operation of a nuclear reactor prototype, and specifies completion dates for the project's two phases, as well as certain other requirements, including licensing. Specifically, EPAct 2005 states that Phase 1 of the project, which requires DOE to, among other things, conduct R&D activities enabling it to select and validate an appropriate technology, culminating in the selection of a technology and initial design parameters, by a target date of September 30, 2011. The law also authorized DOE to submit a report to Congress identifying an alternative date upon which the agency would select the technology and initial design parameters. Phase 2 of the project, in which DOE would develop a final design for a nuclear reactor prototype, apply for licenses to construct and operate the reactor technology, construct the prototype, and begin operations, is to be completed by a target date of September 30, 2021, although DOE is again authorized to submit a report establishing an alternate date for completion. EPAct 2005 also mandated the organization of a consortium of appropriate industrial partners that will carry out cost-shared R&D, as well as design, construction, and operation on behalf of the NGNP Project, and that the NGNP prototype reactor be located at the Idaho National Laboratory in Idaho Falls, Idaho.

In February 2006, the Nuclear Energy Advisory Committee recommended accelerating the NGNP Project schedule to, among other things, make the project more attractive to industry.[23] However, also in February 2006, the administration announced the newly formed Global Nuclear Energy Partnership program, which sought to encourage the expansion of nuclear energy while addressing the burden of spent fuel disposal and the risk of nuclear weapons proliferation and, according to a 2008 National Academy of Sciences report, led to reduced funding for the NGNP Project. Under the Global Nuclear Energy Partnership program, DOE focused advanced reactor R&D activities on developing sodium-cooled fast reactors, and it changed its approach from designing and building a small engineering-scale demonstration of the reactor and reprocessing facility, led by DOE's national laboratories, to accelerating its work with industry to demonstrate commercially viable sodium-cooled fast reactor technology in full-scale facilities.[24]

In 2008, we reviewed NE's Global Nuclear Energy Partnership program and found that DOE's original approach to the domestic component of the program—building engineering-scale facilities—would meet the program's objectives if the advanced spent nuclear fuel recycling technologies on which it focused could be successfully developed and commercialized.[25] However, we also reported that the approach lacked industry participation—potentially reducing the prospects for eventual commercialization of the technologies. NE favored an accelerated approach of building full-scale facilities that would likely require using unproven evolutions of existing technologies that would reduce the long-term benefits of the sodium-cooled fast reactor.[26] We recommended that NE reassess its preference for an accelerated approach. In response, DOE decided, in 2009, to no longer pursue the Global Nuclear Energy Partnership program. However, NE continued research related to sodium-cooled fast reactors, with a new focus on long-term R&D coordinated with NE's fuel cycle research.

Also in 2008, the National Academy of Sciences' National Research Council issued a report reviewing NE's R&D efforts and concluded that the success of any particular advanced reactor technology in the United States would depend on policy decisions and other factors beyond NE's control.[27] In addition, the report concluded that NE's resources were barely adequate for basic studies related to the NGNP Project and entirely inadequate for (1) exploring the sodium-cooled fast reactor at a research level and (2) investigating other reactor technologies. The report also stated that selecting a specific technology to develop from among the options known at that time would have been premature.

In April 2010, NE issued its *Nuclear Energy Research and Development Roadmap: A Report to Congress*, which provides a basis to guide NE's internal programmatic and strategic planning for research going forward. Also in April 2010, NE issued *Next Generation Nuclear Plant: A Report to Congress*, which presents the historical background of the project; details the project's spending; and discusses the principal investments in design, licensing, and research. As discussed in this report, NE selected the high-temperature gas-cooled reactor as the advanced reactor technology to develop under the NGNP Project.

In its 2011 Phase 1 review of the project, the Nuclear Energy Advisory Committee reported that the NGNP Project was not ready to proceed to the complete set of Phase 2 activities, citing

- the need for more detailed design and R&D;
- the need to resolve key licensing issues;
- constraints imposed by EPAct 2005, such as the Idaho site requirement;
- the absence of a public-private partnership, as required;
- absence of industrial partners willing to commit to share in the cost of constructing the prototype reactor; and
- an unrealistic project plan given the limited amount of conceptual design work completed.

The committee recommended that NE continue conducting its Phase 1 R&D, focused on one technology and completing design work, initiate the partnership with the nuclear industry, and continue to engage the Nuclear Regulatory Commission (NRC)—the federal agency that licenses and regulates the nation's civilian use of nuclear reactors—to ensure that the regulatory framework for this new reactor technology would be ready to support commercialization. The committee also recommended eliminating the requirement that the NGNP prototype be sited at Idaho National Laboratory.

In October 2011, DOE submitted a letter to Congress in response to EPAct 2005 requirements to transmit the committee's report and to report on certain requirements to complete Phase 1 of the NGNP Project. As noted above, EPAct 2005 required, by September 30, 2011, that DOE select technology and initial design parameters; alternately, DOE was to submit a report to Congress identifying a new date upon which the agency would do so. In the 2011 letter, DOE stated that it was notifying Congress that the department had not selected the initial design parameters for the NGNP by September 30, 2011, and that it would not proceed with Phase 2 design activities at that time. The letter also stated that DOE would focus on remaining applied R&D, work with NRC on licensing framework, and establish the public-private partnership—in essence, to follow most of the advisory committee's recommendations—until conditions favorable to completing the NGNP Project warranted a change in direction. Further, DOE asserted that the partnership—rather than DOE—would select initial design parameters and would provide an update to the project's schedule and milestones.

In its fiscal year 2014 budget request to Congress, DOE indicated that NE would continue to fund some NGNP research activities under the Advanced Reactor Concepts subprogram.

NE's Approach to Advanced Reactor R&D Supports Several Advanced Reactor Technologies

NE's approach to its advanced reactor R&D is to support research on technologies associated with three main types of advanced reactors: high-temperature gas-cooled reactors; liquid-metal-cooled fast reactors, including the sodium-cooled fast reactor; and fluoride-salt-cooled high-temperature reactors. NE also conducts research supporting other less-developed advanced reactor technologies and supports the development of advanced small modular reactors. NE's approach to advanced reactor R&D addresses broad programmatic goals—including improvements in the economics, safety, and proliferation resistance of nuclear power plants—and aims to develop technologies to reduce nuclear waste and greenhouse gas emissions. This approach provides several advantages, including flexibility in responding to changes in future U.S. energy policy or other circumstances.

NE's Advanced Reactor R&D Efforts Are Focused on Three Primary Reactor Technologies

NE's approach to its advanced reactor R&D efforts primarily is to support research on three main advanced reactor technologies: high-temperature gas-cooled reactors; liquid-metal cooled fast reactors, including the sodium-cooled fast reactor; and fluoride-salt-cooled high-temperature reactors. On a smaller scale, NE also conducts or funds research supporting other less-developed advanced reactor technologies. In addition, NE conducts research supporting a variety of technologies related to the development of advanced small modular reactors. In discussions with representatives from the nuclear industry, members of the National Academy of Sciences' National Research Council, and others, we found that views frequently varied on which specific technologies NE should be supporting through its R&D efforts.

High-Temperature Gas-Cooled Reactors

High-temperature gas-cooled reactors produce energy in the form of high-temperature heat—which can produce electricity or be used as process heat—and differ from existing light water reactors in three key features: using (1) helium gas instead of water as a coolant; (2) graphite instead of water to slow neutrons and sustain the nuclear reaction; and (3) advanced nuclear fuel, which offers safety benefits at high temperatures. These three features make high-temperature gas-cooled reactors capable of operating at higher temperatures than existing light water reactors, thus offering a broader range of applications to industrial processes, as well as higher heat-to-electricity energy conversion efficiencies than are achievable with the lower operating temperatures of light water reactors. The high-temperature gas-cooled reactor is an advanced reactor technology that is expected to be helpful in limiting greenhouse gas emissions, according to NE officials, DOE laboratory staff, and NE documents. In addition, the technology offers inherent and passive safety features—including advanced fuel, helium coolant, and passive heat removal—that are especially important in a post-Fukushima world, according to NE officials.

NE's current involvement with high-temperature gas-cooled reactor R&D began in 2002, when, under the Generation-IV International Forum, high-temperature gas-cooled reactor systems were selected as one of six advanced reactor technologies to be developed by the international consortium. The United States was one of six countries, plus the European Union, that took the lead in developing this reactor technology, according to NE officials. NE chose to pursue high-temperature gas-cooled reactors because, according to

NE officials, laboratory staff, and NE documents, they met the criteria of its advanced reactor programs, including potential improvements over existing light water reactors in safety, economic viability, and their promise for reducing greenhouse gas emissions. NE officials told us that the high-temperature gas-cooled reactor was also chosen because it had operating history in the United States and throughout the world, which meant that many important technical details had already been resolved. At the time, NE concluded that its research would build on existing operational experience, thus increasing the likelihood that high-temperature gas-cooled reactors could be developed and commercialized. NE envisioned its research would further demonstrate the technical and economic viability of the high-temperature gas-cooled reactor technology.

Pursuant to EPAct 2005, NE is to undertake research, development, design, construction, and operation of an advanced nuclear reactor prototype with a targeted completion date of September 30, 2021.[28] From 2005 to 2011, under Phase 1 of the NGNP Project, NE spent more than $500 million on R&D in support of high-temperature gas-cooled reactor technologies. This research was conducted primarily at DOE's national laboratories, and it focused on preliminary reactor design work, developing new reactor fuel, and testing high-temperature materials. The NGNP Project also collaborates with universities and industry on some R&D activities, and NE has coordinated with NRC to conduct R&D necessary to design and license high-temperature gas-cooled reactors in the United States. The last high-temperature gas-cooled reactor was licensed in the United States in 1973, and many questions remain about the process by which NRC will consider license approval for these reactors. According to NE and NRC officials, NE began consulting with NRC in the mid-2000s as NRC worked to (1) update its policy on the regulation of advanced reactors and (2) produce a report to Congress on its advanced reactor licensing strategy for NGNP, which was jointly published in 2008. During this time, NE shared detailed technical information and supporting materials with NRC, information meant to mitigate risks associated with the licensing process, according to NE and NRC officials. Although DOE determined that it would not proceed to Phase 2 design activities for the NGNP Project in 2011, NE continues to fund research on some aspects of a high-temperature gas-cooled reactor, including the testing of advanced reactor fuel and high-temperature materials including graphite.

We found varying views on the potential economic feasibility of the high-temperature gas-cooled reactors. Some industry representatives and a member of the National Academy of Sciences' National Research Council that we interviewed questioned the appropriateness of NE continuing to fund high-temperature gas-cooled reactor research, citing concerns over the reactor's economic viability in relation to the current cost of natural gas. One industry representative cited studies showing that the technology is not economically viable, as well as the fact that NE has, to date, been unable to get industrial partners interested in sharing project development costs. However, an economic analysis by the NGNP Industry Alliance—an international consortium of potential end users, owner-operators, and technology companies brought together to partner with NE and commercialize the high-temperature gas-cooled reactor—concluded that the high-temperature gas-cooled reactor would be economically viable under certain market conditions. In addition, NE has conducted a series of feasibility studies, including detailed economic analyses and studies of potential industrial applications for the process heat, to demonstrate conditions under which the high-temperature gas-cooled reactor technology becomes economically competitive. Representatives from the NGNP Industry Alliance said they believe that, with favorable trends in natural gas prices, the high-temperature gas-cooled reactor will be economically viable by the time the prototype reactor is built in the early-2030s, as is projected under the NGNP Industry Alliance's current project time frames.

Liquid-Metal-Cooled Fast Reactors

The liquid-metal-cooled fast reactors, including sodium-cooled fast reactors, use liquid-metal to cool the reactor core. In addition to producing electricity, a primary benefit of these reactors is in nuclear waste management. Fast reactors can use reprocessed spent nuclear fuel as their energy source, which would help the United States reduce the amount of spent fuel from light water reactors that would need to be stored or eventually placed in a geologic repository. Fast reactors also minimize nuclear waste generation by significantly improving fuel use efficiency as compared to traditional light water reactors.

NE identified the sodium-cooled fast reactor as a reactor technology of interest in 2002 under the Generation-IV International Forum because, according to NE officials, it met criteria of its advanced reactor programs, including significant advances in proliferation resistance, its potential for improving the sustainability of the nuclear fuel cycle, and its management of highly radioactive waste elements. NE continues to conduct research in support of sodium-cooled fast reactors because of their advantages for addressing nuclear waste and because reactors with the same basic technology have been built and operated in the United States and around the world since the 1960s. In fact, several other countries, including Japan and France, had or currently have operating sodium-cooled fast reactors.

We also found varying views on sodium-cooled fast reactors among industry representatives, members of the National Academy of Sciences' National Research Council, and others that we interviewed. Some members of the National Academy of Sciences' National Research Council and an industry representative cited concerns over the safety of sodium-cooled fast reactors—including the highly reactive nature of the sodium in the presence of water and the threat of sodium fires—and believe that these safety issues may never be fully overcome. Moreover, some industry representatives and members of the National Academy of Sciences' National Research Council told us that a fast reactor technology market does not exist in the United States, citing other more cost-effective options for storing spent nuclear fuel, including storage in aboveground casks or water pools as is the current practice at U.S. nuclear power reactors. These individuals believe that NE's sodium-cooled fast reactor research is ill-advised.

In contrast, NE officials and some industry representatives that we interviewed believe that remaining technical challenges with the sodium-cooled fast reactor can be overcome. In addition, NE officials said it is conceivable that changes in government policy for handling spent nuclear fuel in the United States will create a market for fast reactors, as it has in the United Kingdom. Moreover, NE believes that the sodium-cooled fast reactor, or other similar fast reactor technology, may be instrumental in efforts to develop a sustainable nuclear fuel cycle. In 2011, NE undertook a comprehensive study of various options for improving the sustainability of the nuclear fuel cycle, which would potentially entail using fast reactor technology and waste reprocessing to create nuclear power systems that better manage and reduce the generation of nuclear waste when compared to a once-through light water reactor fuel cycle. Although this study did not consider specific advanced reactor technologies, a closed fuel cycle may require using an advanced fast reactor technology, such as the sodium-cooled fast reactor.

Fluoride-Salt-Cooled High-Temperature Reactor

The fluoride-salt-cooled high-temperature reactor design takes advantage of the physical characteristics of liquid-salt coolant to enable the development of a high-temperature system that is scalable to larger power and able to operate at lower pressure and higher power density than the helium-cooled high-temperature gas-cooled reactor. With these characteristics, according to NE officials and DOE documents, these reactors could provide potential safety benefits and increased efficiency over existing light water reactors while maintaining the benefit of providing both electricity and process heat for industrial applications. NE officials told us that this mix of characteristics is the reason why NE provides limited funding research into fluoride-salt-cooled high-temperature reactor. However, the fluoride-salt-cooled high-temperature reactor technology is not very mature or well tested, and because of this is considered more of a long-range advanced reactor technology, according to NE officials. NE funds R&D into fluoride-salt-cooled high-temperature reactors mainly through NE's Nuclear Energy University Program, with the research mainly conducted at universities across the country. Two industry representatives we interviewed took issue with NE for funding a reactor technology that is unproven and that in their view has little chance of ever being built. In response, NE officials told us that the potential benefits of fluoride-salt-cooled high-temperature reactor over other advanced reactor technologies warrant providing limited funds and utilizing university research capabilities.

Other Advanced Nuclear Reactor Technologies

NE conducts or funds R&D on other advanced reactor technologies on a small scale, mainly to assess their potential and better characterize their performance capabilities. NE officials told us that most of these technologies have some unproven aspects, operate in novel ways, or have other characteristics that increase the risk associated with their development. NE supports research into some of these potentially transformative, long-term technology options through its Nuclear Energy University Program. For example, in 2013, the program funded research to assess the feasibility of an advanced reactor fueled with depleted uranium, a design offering a 30-fold increase in uranium ore utilization verses contemporary light water reactor designs.

NE also funds research into promising advanced reactor technologies through the Advanced Reactor Concepts Technical Review Panel process. Through this process, NE identifies R&D needs for potentially viable advanced reactor technologies to inform NE advanced reactor R&D funding decisions. A goal of the process is to facilitate greater engagement between DOE and the nuclear industry. NE first solicited information on advanced reactor proposals from industry in February 2012, after which a review panel made up of experts from national laboratories, universities, and industry reviewed the proposals against established evaluation criteria, including safety, market attractiveness, economics, proliferation risk, waste generation, security, and potential regulatory challenges. The panel's assessment of market attractiveness focused on the proposed technologies' ability to be competitive in the marketplace, and it included variables like efficiency, initial capital costs, and economic factors such as construction, manufacturing, and operating costs and uncertainties, as well as the resulting cost of electricity, according to the Technical Review Panel report. The objective of the Technical Review Panel process was to evaluate the viability of the technologies, understand the R&D needs of each, and prioritize research to support development and commercialization of each. After R&D needs and priorities were identified, NE issued a funding opportunity announcement, competitively selected four projects, and provided a total of $3.5 million in funding for those projects, according to NE officials. Many nuclear industry representatives we interviewed applauded NE's effort and told us that this process was an effective way for NE to collaborate with industry and that it begins to address a long-standing industry concern that NE's R&D efforts did not coordinate with industry or meet industry needs. However, these industry representatives also stated that the $3.5 million in R&D funding was insufficient to meaningfully address the need for collaboration between NE and industry as it was enough to fund a very small number of R&D activities. Notably, Congress provided NE with an additional $12 million to support a continuation of this effort. According to NE officials, NE has issued another industry solicitation and will use the information gathered to make additional industry cost-shared R&D awards early in fiscal year 2015.

Advanced Small Modular Reactors

NE also conducts research on advanced small modular reactors with the goal of supporting the development of innovative small modular reactor designs that offer improved safety, functionality, and affordability. These R&D efforts support advanced small modular reactors that offer simplified operation and maintenance, more efficient energy conversion, and increased proliferation resistance and security. More specifically, NE funds research on advanced sensors, instrumentation and controls, control systems for multiple units, advanced materials, and other major system components. In addition, NE funds efforts to create standards and codes for small modular reactor materials to support the eventual licensing of these advanced reactor technologies.

In 2012 and 2013, through its Small Modular Reactor Licensing Technical Support program, NE issued funding opportunity announcements for cost-sharing with industry for the development of small modular reactor designs—including small modular reactors based on light water reactor technology, as well as advanced small modular reactors—to support the program's vision to provide additional nuclear power options that offer more flexibility in financing, siting, and end-use applications than large light water reactor designs. Under the cost-sharing arrangement for each funding opportunity, DOE is supporting design development, first-of-a-kind engineering, experiments, and analysis in support of gaining design certification approval from NRC for the small modular reactors so that commercial deployment of the first small modular reactor can begin. Industry proposals under these announcements were judged by independent selection panels based on a series of criteria, including the extent to which the design incorporates safety, operability, efficiency, economics, and security characteristics that exceed the capabilities of current reactor designs; the likelihood of expeditiously achieving design certification and deployment;[29] the overall quality of the project plan and business approach; and other factors. Based on our review of the two funding opportunity announcements, however, we found that the funding opportunity announcements differed in the economic information that NE required for proposals. In the 2012 announcement, NE more directly addressed economics and marketability by requiring applicants to propose business plans "to meet expanding domestic electricity requirements at a competitive price" and to "provide their plan to achieve successful commercial deployment" of the technology. By contrast, the 2013 announcement indicated that the economic criteria used to evaluate proposals would be based on the designs' construction, fabrication, deployment, and operational costs. NE officials told us that these criteria indirectly assess economics and marketability of these technologies, and that the type of economic information received from applicants in 2012 was very preliminary and did not provide a good discriminator with which to evaluate proposals. In addition, these officials stated that it was incumbent on the applicants to ultimately assure marketability as they were providing most of the funding and have a profit motive. For both funding opportunity announcements, the panel's evaluation resulted in choosing a small modular reactor design based on conventional light water reactor technology. Some industry representatives, members of the National Academy of Sciences' National Research Council and the Nuclear Energy Advisory Committee, and NE officials told us that this selection was an appropriate choice because it has a significantly better chance of being licensed and constructed in the required time frame, as compared to advanced small modular reactor designs that are not based on conventional light water reactor technology.

NE's Approach Addresses Broad Programmatic Goals and Policy Objectives, and Provides Flexibility in Responding to Changes in U.S. Energy Policy

While the broad goals of NE's advanced reactor R&D efforts are to improve the economics, safety, and proliferation resistance of nuclear power plants, the R&D efforts also aim to develop advanced reactor technologies that can prepare the United States to address policy objectives such as reducing nuclear waste and greenhouse gas emissions. NE's approach to advanced reactor R&D is to conduct research in support of multiple advanced reactor technologies. According to NE officials and documents, because NE's approach to advanced reactor R&D has multiple goals and seeks to address several different policy objectives, NE works on multiple technologies simultaneously. A key objective of NE's advanced reactor R&D efforts is to conduct research to remove technology barriers or reduce technology risks, while collaborating with industry and academia, with the ultimate goal for industry to take the results of NE's research to the next step of development and commercialization. NE focuses on R&D that industry does not have the means to carry out, according to NE officials, with the expectation that the research will reduce financial risks to industry and thereby increase the affordability of industry investment in new nuclear technologies. In addition, NE engages and collaborates with NRC on issues related to the eventual licensing of advanced reactors, including understanding the likely scope and extent of R&D necessary to support the licensing process.

While advanced reactors are attractive for many reasons, NE carries out research on a variety of reactors because, in part, different reactor types can address particular objectives, according to NE officials. For instance, fast reactors can be better at addressing the nuclear waste issue than some other advanced reactors, while high-temperature gas-cooled reactors provide process heat and may be a better solution for addressing greenhouse gas emissions. According to NE officials, the development of fast reactors, such as the sodium-cooled fast reactor, is likely to play a critical role in managing spent nuclear fuel if and when the United States decides to reprocess and use its spent nuclear fuel rather than store it at reactor sites,[30] as is the current practice at U.S. nuclear power reactors, or isolate it in a geologic repository underground, as has been proposed.[31]

To remain aware of industry's R&D needs and international nuclear energy developments, NE regularly collaborates with industry and international organizations, according to NE officials and NE documents. NE officials told us that NE regularly collaborates with industry on specific R&D projects by sharing technical data and information. For example, NE is currently collaborating with industry on advanced fuels and materials, among other things. NE officials told us they work with industry and have conversations regarding specific R&D activities. According to NE officials and some industry representatives that we interviewed, this type of collaboration has been increasing in recent years, and one industry representative stated that such collaboration is critically important to ensuring that NE's activities are relevant for industry.

One way that NE has recently collaborated with industry, according to NE officials, was through the Advanced Reactor Concepts Technical Review Panel process. Some industry representatives we talked to stated that this review panel process was beneficial to both industry and NE because it helped inform NE of industry R&D needs and because it has provided some funds to industry to carry out research on promising new technologies. Industry officials also told us that the process has opened up some new channels of communication between NE and industry. However, industry representatives also stated that, although this collaboration with industry is beneficial, NE could be doing more to ensure that its R&D is more fully aligned with industry needs. For example, according to one industry official, NE conducts some research that, while interesting and potentially beneficial, has little utility for industry's current needs.

NE carries out international collaboration through ongoing meetings of the Generation-IV International Forum and through the International Atomic Energy Agency,[32] through the Organisation for Economic Co-operation and Development (OECD),[33]and through bi-lateral agreements with many countries around the world, including Canada, the Russian Federation, the People's Republic of China, Japan, the Republic of Korea, and countries in the European Union. NE officials cited several examples of such collaboration, such as with the People's Republic of China on high-temperature gas-cooled reactors; with France on their sodium-cooled fast reactor development project; with Japan on advanced materials for sodium-cooled fast reactors; and with the Republic of Korea on sodium-cooled fast reactors.

NE's approach to advanced reactor R&D provides several advantages, primarily flexibility in responding to changes in future U.S. energy policy or other circumstances, according to NE officials. The officials said they believe that conducting research in support of multiple advanced reactor technologies gives the agency the flexibility to respond to external factors affecting the direction of their advanced reactor R&D efforts, including changes in U.S. energy policy, energy markets, or other areas. Specifically, NE officials told us that the current approach positions NE to respond to changes in U.S. energy policies, such as policies for managing the nation's nuclear waste or controlling greenhouse gas emissions. Changes in either of these policies would affect the direction of NE's advanced reactor efforts, according to NE officials. For instance, a policy calling for the United States to manage nuclear waste by reprocessing spent nuclear fuel and reusing it as reactor fuel would result in NE focusing efforts and concentrating resources on developing and deploying fast reactor technologies. NE officials told us that the current practice of storing nuclear waste in aboveground facilities at nuclear power plants across the country will eventually be changed, and waste will either be moved to long-term underground repositories, or reprocessed and burned in fast nuclear reactors.[34] NE officials told us that ongoing research into fast reactor technologies is important so that NE is positioned to react to changes in U.S. policy toward the handling of nuclear waste, including waste that has already been generated and waste that continues to be generated.

Similarly, NE officials said that policies that affect the prices of various energy sources would have an effect on the commercial attractiveness of high-temperature reactors, including high-temperature gas-cooled reactors. For instance, officials cited the possibility of the imposition of a carbon tax to control greenhouse gas emissions, or the possibility of natural gas prices rising, either of which would make nuclear energy more economically competitive and increase the attractiveness of the high-temperature gas-cooled reactors that produce both electricity and process heat for industrial applications. These industrial applications currently rely heavily on natural gas or coal plants as their source for high-temperature process heat. NE officials stated that natural gas prices in other countries are already at levels where the high-temperature gas-cooled reactors are projected to be economically competitive and that this has resulted in some interest from outside the United States in the development of this technology

According to NE officials, another advantage of NE's approach to advanced reactor R&D is that NE is able to maintain an employee base with knowledge and expertise on a wide variety of reactor technologies. NE officials told us that maintaining staff expertise is important so NE can continue to conduct research on the various technologies, train the next generation of scientists and engineers, and be ready to support the production of prototype reactors when the time comes. In addition, maintaining a level of expertise in a variety of advanced reactor technologies means that NE can engage with, monitor, and support other countries as they develop advanced reactor technologies. These officials said this is important because other countries are actively developing advanced reactor technologies, and the United States needs scientists that can understand how those reactors operate, in part, to judge their safety and nuclear proliferation risks.

NE officials also said that conducing R&D on several types of advanced reactors simultaneously, rather than focusing on a single reactor type, also gives NE the ability to fund R&D supporting promising but unproven reactor technologies. For instance, NE is funding limited research on lead-cooled fast reactors, which offer the potential for improved safety and proliferation resistance over other advanced reactor technologies, but they have some unproven technologies and components, according to NE officials. Similarly, NE is funding research in support of an advanced small modular reactor based on fast reactor technology that would potentially address the nuclear waste issue and also provide process heat for industrial applications. NE officials said that it is important to have funds available to support these and other potentially game-changing technological breakthroughs. However, in its June 2013 report, the Nuclear Energy Advisory Committee was critical of NE's approach, saying that NE needs to better prioritize its R&D efforts on a smaller number of advanced reactor technologies to focus research funding on the ultimate goal of deploying an advanced reactor prototype.

Although NE selected the technology to develop under the NGNP Project, many members of the National Academy of Sciences' National Research Council, members of the Nuclear Energy Advisory Committee, and industry representatives we interviewed agree with NE's approach to advanced reactor R&D because the time is not right for NE to move to the deployment phase. For instance, representatives from industry and the Nuclear Energy Advisory Committee told us that uncertainties around current policies for handing nuclear waste and controlling greenhouse gases do not make a compelling case for choosing an advanced reactor technology to deploy as a prototype. The 2008 National Academy of Sciences' National Research Council review of NE's advanced reactor R&D efforts agreed with NE's approach to advanced reactor R&D, saying that there are several policy matters and other questions—undetermined nuclear waste management options, unformulated environmental policy, ongoing work of other countries on advanced technologies, and unclear nonproliferation regimes—that will affect NE's decisions and priorities. This review team stated that, given these unknowns, it would be premature to select a winning technology from among current options.

In addition, in January 2012, the President's Blue Ribbon Commission on America's Nuclear Future recommended having the United States continue multiple near-term (i.e., light water reactor) and long-term (e.g., small modular reactor, sodium-cooled fast reactor, high-temperature gas-cooled reactor) R&D efforts until NE could defensibly select technologies that would meet certain regulatory and policy requirements (e.g., safety, environmental protection, security, and nonproliferation). Moreover, members of the Nuclear Energy Advisory Committee and the National Academy of Sciences' National Research Council, and representatives from industry told us that current NE funding levels would prohibit NE from deploying a prototype reactor even if NE chose an advanced technology to deploy. Some of them said that NE is correctly positioning itself to be prepared to deploy a prototype reactor in the long-term as policies or energy markets change. One Nuclear Energy Advisory Committee representative said that United States could focus its advanced reactor R&D efforts quickly in response to a policy change or other congressional direction, provided that NE also saw increased funding.

NE Uses Internal and External Reviews to Set Program and Funding Priorities but Does Not Have a Strategy for Deploying an Advanced Reactor Prototype

NE's uses internal and external reviews to set program and funding priorities for advanced reactor R&D and to evaluate progress toward program goals. However, NE does not have a strategy for overcoming barriers to deploying an advanced nuclear reactor prototype, increasing the likelihood that such a reactor will not be built by the 2021 target date specified in EPAct 2005. Not deploying a prototype carries certain risks, including waning U.S. influence in the safe operation of nuclear plants internationally and potential loss of certain knowledge and expertise.

NE Plans, Prioritizes, and Evaluates Its R&D Activities through Internal and External Reviews

NE takes a number of steps to plan and prioritize its advanced reactor R&D efforts and evaluate progress toward program goals. Before its annual program planning meetings, NE and national laboratory staff develop a list of R&D efforts considered to be priorities. NE management reviews this information in light of program goals, including long-term goals described in NE's 2010 R&D Roadmap, program funding, and schedules, according to NE officials. Once the research priorities are established and approved by management, the individual laboratories develop detailed work plans, which describe the objectives and scope of the work to be performed. These work plans are reviewed to ensure that the proposed work is aligned with NE's mission and that the work can be accomplished within the allotted budget and time frames, according to NE officials. All of the approved work plans are then entered into NE's performance management system—the Program Information Collection System—which allows NE to track progress toward budget and schedule milestones on an ongoing basis. According to a laboratory staff member, this system tracks progress toward short-term goals—on a monthly basis—and long-term goals—on yearly, 3-year, and 5-year time frames.

NE monitors and evaluates its advanced reactor R&D activities on an ongoing basis through the Program Information Collection System and conducts program reviews—monthly, quarterly, and annually—to assess progress toward program goals, according to NE officials. For example, officials from the Advanced Reactor Concepts and Advanced Small Modular Reactor subprograms hold monthly progress review meetings to discuss, among other things, program updates, technical highlights, and budget and milestone status updates. The officials use monthly status tracking reports generated by the performance management system as part of these reviews, in which officials review cost and schedule performance data. In addition to the monthly meetings, officials from the Advanced Reactor Concepts and Advanced Small Modular Reactor subprograms typically conduct four in-depth reviews of each year, according to NE officials. These reviews focus on one or more specific areas of research, and officials discuss progress toward goals, important issues or problems, and plans going forward. For example, the meeting minutes from the quarterly review of the Advanced Reactor Concepts and Advanced Small Modular Reactor subprograms in July 2013 show that officials discussed accomplishments and also priorities for the upcoming fiscal year, and conducted in-depth discussions of certain program areas and overviews of others. In addition, NE conducts annual reviews of activities across multiple subprograms and topics to ensure that NE's efforts are complementary and nonduplicative, and also to gain insight into areas of potential collaboration. For example, during its annual review of the nuclear reactor R&D efforts in March 2014, NE officials discussed progress on fuels for the high-temperature gas-cooled reactor, advanced reactor licensing, and advanced reactor materials for small modular reactors, among other things. On a less-formal basis, management officials at the national laboratories are in frequent communication with NE management through weekly teleconferences to provide regular progress updates and provide information on unforeseen circumstances or challenges, according to NE officials.

NE also takes steps to coordinate efforts across its R&D programs and subprograms to leverage experience and funding, as well as to reduce redundant R&D activities. Officials from the Advanced Reactor Concepts and Fuel Cycle subprograms stated that they frequently coordinate with each other because their R&D efforts are interdependent. For example, the Fuel Cycle subprogram is conducting R&D on accident tolerant fuels that will be used for advanced reactors, so coordination between the Fuel Cycle subprogram and the Advanced Reactor Concepts subprogram is crucial, according to NE officials.

To further help ensure that R&D efforts are coordinated and to minimize redundancies, NE established the Nuclear Energy Enabling Technologies program in fiscal year 2011. The program is designed to conduct R&D on crosscutting technologies that complement NE's activities to support and enable the development of new advanced reactor designs and fuel cycle technologies. NE created the program to better coordinate and integrate R&D activities after officials identified some similar efforts being performed on crosscutting areas, such as materials, across more than one program, according to NE officials. Through this program, NE has awarded over $9 million to support R&D projects focused on reactor materials, advanced sensors and instrumentation, and advanced methods for manufacturing, among other things. NE determines which R&D efforts are conducted by the Nuclear Energy Enabling Technologies program by reviewing R&D proposals submitted by different groups, including the national labs, universities, research institutions, and industry, according to NE documents. Specific R&D projects are selected based on common needs of programs and subprograms, with each selected project required to support at least three programs or subprograms.

Others also periodically conduct external reviews of NE's advanced reactor R&D to inform the planning and prioritization efforts for and assess the progress of its R&D activities. Most prominently, according to NE officials, the Nuclear Energy Advisory Committee provides NE with independent advice and recommendations on complex science and technical issues that arise in planning, managing, and implementing NE's R&D activities. The Nuclear Energy Advisory Committee typically meets twice annually with NE management to discuss its reports and recommendations. The Nuclear Energy Advisory Committee's subcommittee on Nuclear Reactor Technology is intended to provide expert guidance to NE on both the short-term and long-term direction of its R&D efforts on reactor technologies. NE officials and Nuclear Energy Advisory Committee representatives told us that the committee has been beneficial in providing expertise to NE and that NE has been responsive to the committee's recommendations. In 2011, a Nuclear Energy Advisory Committee review of NE's R&D efforts on the NGNP Project determined that NE should not move forward with the complete set of Phase 2 activities of the project, citing constraints imposed by EPAct 2005 and difficulties finding industry partners. Subsequently, in 2011, NE informed Congress that it would not proceed with Phase 2 design activities of the NGNP Project until circumstances warranted a change in direction.

NE's efforts have also been reviewed by other outside entities, including the Secretary of Energy's Advisory Board, which provides advice and recommendations to the Secretary of Energy on various topics. For example, in 2012, the Secretary of Energy requested the board identify areas in which standards for safety, security, and nonproliferation should be developed for small modular reactors; identify challenges, uncertainties, and risks to commercialization; and provide advice on approaches to manage these risks and accelerate deployment of these reactors. The board determined that the commercialization of small modular reactors was likely to produce multiple benefits for the country, including helping provide for a more reliable power grid with more widely distributed power generation once current light water reactors are retired; supporting clean generation and reduced carbon emissions; and helping preserve influence of the United States on nuclear nonproliferation issues. The board stated that to deploy small modular reactors widely, the nation must develop a robust small modular reactor industry that can manufacture cost-competitive small modular reactors that meet U.S. regulatory standards, and that the primary risk for commercialization of these reactors, beyond design certification and licensing, is the ability to drive the plant costs down sufficiently to become competitive with other energy sources, such as natural gas, without compromising safety and security. To develop this industry, according to the board, the U.S. government will likely have to play a significant financial role beyond the Small Modular Reactor Licensing Technical Support program.

NE Does Not Have a Strategy to Overcome Several Barriers to Deploying an Advanced Nuclear Reactor Prototype

Although NE's primary mission is to advance nuclear power through research, development, and demonstration, its deployment of an advanced reactor prototype under the NGNP Project is unlikely in the foreseeable future. Among the different advanced reactor technologies currently supported by NE R&D, the high-temperature gas-cooled reactor technology is the most likely to be deployed and commercialized in the near term, according to an NE planning document. NE officials said that the likelihood is based on the wide range of potential market applications to industry of electricity and process heat and is supported by substantial government investments in the technology's development, including testing of materials, fuels, and other components. NE has consulted with the NGNP Industry Alliance on the project, including discussing the alliance's plan for proceeding with development of a prototype reactor. In addition, NE has done market research on potential industrial applications for the process heat. Further, NE established a contract in 2013 with NGNP Industry Alliance to develop economic analyses detailing how industry may best engage in developing and commercializing high-temperature gas-cooled reactor technologies. According to laboratory staff, development and testing of the advanced fuel for high-temperature gas-cooled reactor has progressed positively, and other research on high-temperature materials and other components has produced positive results.[35]

In 2011, DOE informed Congress that it would not proceed with Phase 2 design activities for the NGNP Project until circumstances warranted a change in direction. According to NE officials, laboratory staff, and representatives of the NGNP Industry Alliance, the NGNP Project remains hindered by several barriers. Specifically, barriers are as follows:

- **Cost-share requirements.** DOE's attempts to implement the cost-share provisions in EPAct 2005 for the NGNP Projects have met with resistance from industry, according to DOE officials and industry representatives, because of differences in how EPAct 2005 is interpreted by NE and by the NGNP Industry Alliance. EPAct 2005 provides that activities by industry must be cost-shared in accordance with the research, development, demonstration, and commercial application cost-sharing provisions established under section 988 of the act. Specifically, the Secretary must require cost-sharing in accordance with this cost-sharing provision when carrying out a research, development, demonstration, or commercial application program or activity that is initiated after August 8, 2005. For applied research and development activities, industry generally is to provide not less than 20 percent of the cost, but the cost-share may be reduced or eliminated if the Secretary determines doing so is necessary and appropriate. For demonstration and commercial application activities, industry generally is to provide not less than 50 percent of the cost, but the cost-share may be reduced if the Secretary determines that doing so is necessary and appropriate considering any technological risk relating to the activity. However, according to NE officials and representatives from the NGNP Industry Alliance, they have been unable to come to an agreement on implementing the cost-share requirement for funding the remainder of the NGNP Project because of disagreement on the applicable cost-share levels and how and when the cost-share would be applied to specific activities or project phases. The NGNP Industry Alliance favors meeting the total cost-share requirement by measuring costs over the course of the remainder of the NGNP Project rather than on an annual basis. According to the NGNP Industry Alliance cost-share proposal from November 2009, the alliance suggested assessing a lower industry cost-share in the first years of the project and increasing the industry share over time, with industry paying the vast share of the annual project costs by the final years of the project. Under this proposed scenario, the alliance states that the cumulative industry contribution would meet the overall cost-share requirement, and NE's portion of the development costs would largely be paid up front. DOE did not fully consider the alliance's proposal for a multiyear approach to the cost-share requirement, according to NE officials, because the project was not proceeding at the time due to funding constraints, competing program priorities, and other factors

Representatives from the NGNP Industry Alliance told us that cost-sharing the development activities on a annual basis is not feasible because it would mean a substantial layout of funds with a very long payoff time and would expose the industry partners to significant financial risks. NGNP Industry Alliance representatives said these risks include unknowns associated with obtaining regulatory approval from NRC for the prototype reactor, and the risk that NE will not be provided sufficient funds through congressional appropriations to meet its obligations. NE officials told us that they understand the NGNP Industry Alliance's perspective and had been attempting to work out an agreement when DOE decided not to proceed to Phase 2 of the project. Representatives from the NGNP Industry Alliance told us that the impasse over cost-sharing needs to be resolved in order to proceed with the NGNP advanced reactor prototype.

- **Site requirement.** According to NE officials, laboratory staff, and NGNP Industry Alliance representatives, the EPAct 2005 requirement that the NGNP reactor prototype be located at the Idaho National Laboratory is another barrier to proceeding with the project. Representatives of the NGNP Industry Alliance said that part of the economic benefit of the reactor prototype would be the use of the high-temperature process heat that results from operating the high-temperature gas-cooled reactor. Alliance representatives said that building the reactor at Idaho National Laboratory foregoes the economic benefit because industries that could potentially use process heat are not located near the laboratory, making the overall prototype reactor less economically attractive. Instead, they told us that the NGNP prototype reactor should be located where the petrochemical or other industries that use process heat could benefit from it. This is consistent with a finding in the Nuclear Energy Advisory Committee's 2011 Phase 1 review of the NGNP Project. In its review, the committee stated that the business case to optimize NGNP use for process heat applications and electricity indicates that a site in proximity to a wide range of industrial uses is more appropriate and that a siting at the Idaho National Laboratory will not support a partnership agreement with industry. If industry cannot realize an economic benefit from the prototype reactor, it is unlikely that industry would support the reactor being built at the Idaho National Laboratory.
- **Fiscal constraints and competing priorities.** NE officials, laboratory staff, industry representatives, and Nuclear Energy Advisory Committee members that we interviewed told us that NE's recent

funding levels are inadequate to move forward with the NGNP prototype reactor. NE officials and the NGNP Industry Alliance both estimate that NE's share of NGNP Project could amount to as much as $2 billion over the remainder of the project, in which costs would be shared with industry. Under the NGNP Industry Alliance's proposal, DOE would provide between approximately $170 million and $330 million annually over the first 6 years of the proposed plan. This compares to the total funding for Advanced Reactor Concepts subprogram of about $60 million in 2014. NE officials, Nuclear Energy Advisory Committee members, and laboratory staff told us that NE's funding levels are inadequate to move forward toward a prototype reactor, even if it were to focus its resources on one effort. Furthermore, an NE officials and a member of the Nuclear Energy Advisory Committee told us that current priorities to fund R&D aimed at sustaining the existing light water reactors and focusing on the design and licensing of small modular reactors would have to shift in order to make more funds available for advanced reactor R&D.

- **Competition from natural gas.** NE officials, some industry representatives, and Nuclear Energy Advisory Committee members that we interviewed told us that low natural gas prices have made nuclear energy less attractive economically over the past few years, reducing overall interest in nuclear power options. NE officials and industry representatives said that the current atmosphere is not conducive to partnering with industry on advanced reactor projects, including the NGNP Project.

The Secretary's October 2011 letter to Congress did not specify which conditions might warrant a change in program direction—that is, proceed with Phase 2 of the NGNP Project—and NE has not developed a strategy for overcoming the identified barriers hindering the resumption of the project or a set of criteria for determining when a change in program direction would occur. An NE management official that we interviewed stated that conditions that would warrant a change in direction might include Congress legislating a carbon tax, a rise in price of natural gas, or an increase in funding for the NGNP. In addition, developing such a strategy may involve consultation with the Nuclear Energy Advisory Committee and others, including independent nuclear experts. Without a strategy for overcoming the barriers hindering the restart of the NGNP Project and identifying conditions NE can use for determining when a change in program direction would occur, it will be difficult for NE to demonstrate that it is poised to move forward, and it risks the project being on hold indefinitely.

According to EPAct 2005, NE was required to select the initial reactor design parameters to be used for the NGNP Project by September 30, 2011, or submit a report to Congress establishing an alternative date for making the selection. However, the Secretary's 2011 letter to Congress did not specify initial design parameters for the NGNP or specify an alternative date for making a selection. Instead, the letter stated that the initial design parameters had not yet been selected and that such a selection would be made by the public-private partnership once it is formed. Without selecting initial reactor design parameters or establishing a date to make a selection as required by EPAct 2005, it is not clear if or when NE is going to take this next step in deploying the NGNP prototype reactor.

In addition, after the Secretary's decision not to proceed with Phase 2 design activities, NE's engagement with NRC on licensing issues has decreased. NE maintains a team seeking to engage NRC on NGNP licensing issues, according to NE officials, but NRC has reassigned staff from its NGNP work and engages with NE on a minimal basis, according to NRC officials. NRC officials said that they cannot proceed substantively further in developing a licensing framework until NE has developed a specific design for an advanced reactor technology.

Furthermore, not deploying an advanced reactor prototype carries some risks, according to some industry representatives, Nuclear Energy Advisory Committee members, and NE officials we interviewed. Specifically, these risks include (1) falling behind other countries in advanced reactor development and losing market share in the global market for nuclear energy; (2) losing influence on which reactor technologies are developed, which raises safety and nuclear proliferation concerns; and (3) losing its ability to manufacture the components necessary to construct nuclear plants.

By not deploying an advanced reactor prototype, the United States risks falling behind other countries—such as Japan, Russia, China, South Korea, and France—that are actively working to deploy and commercialize advanced reactors, according to a Nuclear Energy Advisory Committee report. For example, Russia currently has two sodium-cooled fast reactors—one experimental and one commercial—in operation and is developing or constructing additional sodium-cooled fast reactor technologies, and it has plans to export reactor technology to other nations. In addition, China has an operating sodium-cooled fast reactor and high-temperature gas-cooled reactor on a test reactor scale and is in the process of building a prototype high temperature gas-cooled reactor, according to NE officials. In potentially losing its global leadership position in developing nuclear technologies, the United States risks losing market share in the global market for nuclear energy, which would cost the U.S. economy high-paying jobs in the nuclear industry, according to these individuals.

By falling behind other countries in advanced reactor development, the United States also risks losing influence on determinations of which reactor technologies are developed, with implications for the safety of reactor operations worldwide, as well as implications for how resistant the technologies are to nuclear proliferation—including the safe, effective disposal of nuclear waste—according to NE officials, laboratory staff, and Nuclear Energy Advisory Committee members that we spoke to. Specifically, according to members of the Nuclear Energy Advisory Committee, if the nation is not leading the development of advanced reactors, other countries may operate reactors that do not meet the highest safety standards and may not take adequate steps to ensure nuclear waste is handled appropriately and properly secured.

Similarly, the United States risks losing its ability to manufacture the components necessary to construct nuclear plants, according laboratory staff and Nuclear Energy Advisory Committee members that we interviewed. In addition, by not deploying an advanced reactor, NE risks losing staff, including engineers with the knowledge and experience necessary to design and build advanced reactors, according to an NE official and laboratory staff. For example, without a specific goal of developing an advanced reactor prototype, NE staff are more likely to leave NE for jobs with a better sense of mission, according to these officials.

CONCLUSION

Energy demand in the United States is expected to rise considerably over the coming decades, and concerns remain over energy security and greenhouse gas emissions from the burning of fossil fuels. While nuclear energy accounts for about 20 percent of electricity generation in the United States and produces no air pollution or greenhouse gases, the accident at Japan's Fukushima Daiichi commercial nuclear power plant in March 2011 highlighted ongoing concerns about the safety of nuclear plants, and concerns also exist about potential threats of nuclear proliferation and terrorism. With this in mind, it is important that nuclear power plants continue to evolve and provide energy economically, while also addressing safety and proliferation concerns. By conducting nuclear reactor R&D, NE has a critical role to play as it supports existing light water reactors, as well as a new generation of advanced nuclear reactors.

However, in 2011, DOE informed Congress that it would not proceed with Phase 2 of the NGNP Project until circumstances warranted a change in direction, and the project remains hindered by several barriers, including the cost-share and site requirements of EPAct 2005. NE officials have attempted to work out a cost-share agreement with the alliance, but different interpretations of the cost-share requirements by DOE and the NGNP Industry Alliance have created an impasse, and no agreement had been reached before DOE determined that it would not proceed to Phase 2 of the project.

Another barrier to proceeding with the project is the EPAct 2005 requirement that the NGNP reactor prototype be located at the Idaho National Laboratory. Building the reactor there foregoes the economic benefit of the reactor's process heat because industries that could potentially use the prototype reactor's high-temperature process heat are not located near the laboratory. If industry cannot realize an economic benefit from the prototype reactor, it is unlikely that industry would support the reactor being built at the Idaho National Laboratory.

Moreover, DOE's October 2011 letter notified Congress that the department had selected the NGNP technology, as required by EPAct 2005, acknowledged that the department had not selected the initial design parameters for the NGNP or identified the date upon which it would do so by September 30, 2011, as required by EPAct 2005, and essentially put Phase 2 of the NGNP Project on hold until conditions favorable to completing the NGNP warranted a change in direction. However, the letter did not specify which conditions might warrant a change in program direction—that is, proceed with Phase 2 of the NGNP Project—and NE has not developed a strategy for overcoming the identified barriers hindering restarting the project or which contains such conditions. Without a strategy for overcoming the barriers hindering the restart of the NGNP Project and identifying conditions NE can use for determining when a change in program direction would occur, the project may be on hold indefinitely. Furthermore, without selecting initial reactor design parameters and reporting the parameters to Congress, as required by EPAct 2005 for completing Phase 1 of the project, or establishing a date to make a selection, it is not clear if or when NE is going to take this next step and proceed with Phase 2 of the NGNP Project.

Recommendations for Executive Action

To better prepare the Department of Energy to meet the requirement of the Energy Policy Act of 2005 to deploy the NGNP prototype reactor, we recommend that DOE take the following two actions:

- Develop, in consultation with the Nuclear Energy Advisory Committee and independent nuclear experts, as appropriate, a strategy to proceed with Phase 2 of the NGNP Project, outlining conditions that will warrant a change in program direction, remaining research and development activities, projected project budget and schedule, and steps necessary to overcome barriers to successful completion of the NGNP Project.
- Provide a report to Congress complying with the statutory requirement for Phase 1 of the NGNP Project by providing initial design parameters or a date for providing them. The report should also provide an updated status of the issues DOE identified in its 2011 letter to Congress and outline any additional barriers to proceeding with Phase 2 of the project, including
  - the status of the establishment of a public-private partnership;
  - the project strategy, including conditions that would warrant restarting the project; and
  - a legislative proposal, if necessary, to address any barriers to proceeding with the project, including the site and cost-share requirements.

Agency Comments and Our Evaluation

We provided a draft of this report to DOE for review and comment. In written comments, DOE's Assistant Secretary for Nuclear Energy, responding on behalf of DOE, wrote that DOE agreed in principle with our first recommendation and respectfully disagreed with our second recommendation. In addition, DOE provided technical comments, which we incorporated in the report as appropriate.

In its comment letter, DOE stated that it agreed in principle with our recommendation that it develop a strategy to proceed with Phase 2 of the NGNP Project. Moreover, DOE stated that its current strategy is to continue updating analyses of requirements for successful commercialization of reactor technology to reflect changing market conditions, research and development accomplishments, and the maturity of the licensing framework. However, this strategy does not outline steps DOE could take to proactively overcome the barriers hindering the resumption of the NGNP Project, nor does it outline criteria for determining when a change in program direction would occur. We continue to believe that developing a strategy to proceed with Phase 2 of the NGNP Project is important because without a strategy it will be difficult for NE to demonstrate that, upon completion of Phase 1, it will be poised to develop a final design and construct and operate the prototype reactor. Moreover, not having a strategy for proceeding with Phase 2 could result in the project being on hold indefinitely.

DOE respectfully disagreed with our recommendation that it provide a report to Congress that, among other things, provides initial design parameters or a date for providing them and outlines barriers to proceeding with Phase 2 of the project. DOE stated that such a report was not advisable or useful, or necessary as a means for the Department to comply with the statutory requirements for Phase 1 of the NGNP Project, and further stated that the Department is in compliance with the relevant statutory requirements. As DOE explained, it reported to Congress in 2011 that while it had selected the hydrogen production technology, as required by EPAct 2005, it had not selected the initial design parameters for the project and that based on the recommendations of the Nuclear Energy Advisory Committee, fiscal constraints, competing priorities, projected cost of the prototype, and the inability to reach agreement with industry on cost sharing, it would not proceed with Phase 2 design activities at that time. Instead, it would continue to focus on high-temperature reactor R&D activities and establishment of a public-private partnership, among other things, until conditions warranted a change in direction. DOE did not, however, establish an alternative date for selecting the initial design parameters, as EPAct 2005 required. Rather, it stated that selection of initial design parameters would be made by the public-private partnership once it is formed. Given that almost 3 years have passed since the letter to Congress, we believe that the recommended report is warranted and would serve to inform Congress of the status of the NGNP Project and provide transparency and accountability regarding DOE's intentions for completing Phase 1 and proceeding with Phase 2 of the project. For example, by providing a firm date for selecting the initial design parameters of the NGNP prototype reactor, DOE could be held accountable to meeting that date or could engage in a discussion about whether and why that date should be further extended. Similarly, an updated report to Congress could include a candid description of the ongoing barriers to moving forward, which could spur discussions resulting in legislation or other remedies to mitigate these barriers.

Sincerely yours,

Frank Rusco
Director, Natural Resources and Environment

Timothy M. Persons, Ph.D., Chief Scientist
Director, Center for Science, Technology, and Engineering

Appendix I: Overview of Nuclear Reactor Technologies

Overview of Reactor Operations

Nuclear reactors generate heat by sustaining a fission chain reaction in nuclear fuel. Nuclear fission reactions can occur when a neutron strikes the nucleus of a large atom, causing that nucleus to split, or fission. The result of a fission reaction is typically two fission fragments, or smaller nuclei; 2 or 3 new fast-moving neutrons; and significant heat. In a nuclear reactor, the large atoms used for fission are typically the fissile isotopes uranium-235 or plutonium-239, and the new neutrons produced by a fission reaction are used to initiate new fission reactions, resulting in a sustained fission chain reaction.[36] This heat generated by this fission reaction is typically used to create steam and drive a steam turbine to generate electricity. Some reactors may also operate at particularly high temperatures and can use the heat to either generate electricity or to supply process heat that can be used for various industrial processes, replacing other heat sources such as natural gas.

Types of Nuclear Reactors

Nuclear reactors typically fall into one of two types based on the neutron spectrum, or neutron energies, at which the fission reactions occur as follows:

- Thermal reactors optimize the fission reaction rate in their fuel. This is done by slowing down, or moderating, the high-energy fast neutrons that are the products of fission reactions. This thermalization of the fast neutrons increases the likelihood that a neutron will initiate a fission reaction. Currently deployed light water reactors, including pressurized water reactors and boiling water reactors, are thermal reactors.
- Fast reactors, by contrast, do not moderate the fission neutrons, leaving them as fast neutrons. A fast neutron has a lower likelihood of initiating a fission event than a slow neutron, so the chain reaction can be more difficult to sustain, but it has the benefit of producing more neutrons when fission does occur. These surplus neutrons, as compared to the number of neutrons produced in thermal reactors, allow fast reactors to be more effective than thermal reactors at creating, or breeding, new fuel through neutron bombardment of uranium-238 (creating plutonium-239). Fast reactors optimized for fuel production in this manner are called fast breeder reactors and can produce more fuel through breeding than they consume. Fast reactors may also use spent fuel from other nuclear reactors as fuel and thereby reduce long-term fuel disposal needs.

Overview of the Reactor Core

While there are a large number of reactor technologies that can differ significantly, the fission reaction in a reactor occurs in the central region of a reactor called the reactor core. The reactor core typically contains several components as follows:

- **Nuclear fuel.** Nuclear reactors need fissile isotopes, such as uranium-235 and plutonium-239, to sustain chain reactions. Commercial reactors often use uranium ore that has been enriched in the isotope uranium-235 as their fissile fuel; the rest of the fuel consists of the non-fissile uranium-238. However, reactor operation will result in the conversion of some uranium-238 to the fissile isotope plutonium-239, which may then fission and contribute to power generation, and some reactor fuel may start with some of the uranium-235 mixed with plutonium-239 (sometimes referred to as a mixed oxide fuel). Fast reactors can also use spent fuel from other reactors as fuel and can be very effective at converting uranium-238 into plutonium-239. Some reactors can also utilize uranium-233 or thorium-232 as components of their fuel.

- **Moderator.** Thermal reactors use a moderator material to slow down, or thermalize, the fission neutrons in order to sustain the fission reaction. This is needed because neutrons produced by fission reactions are too fast (or energetic) to have a high likelihood of initiating a new fission reaction in fuel. Fast reactors are designed to utilize fast neutrons for the fission reactions and fuel breeding and, as such, do not use a moderator.
- **Coolant.** To remove heat from the core, a coolant—typically water, a gas, or liquid metal—is circulated through the core. The coolant both prevents the core from overheating (which could damage or melt the fuel) and it carries energy, in the form of heat, outside the core for electricity production, typically by generating steam that then drives a steam turbine. In some reactor types, the coolant can also function as the reactor's moderator.
- **Reaction control.** Reactors can use different techniques to maintain the fission chain reaction at appropriate rates. For example, control rods may be inserted into reactor cores to absorb neutrons and slow down (or stop) the chain reaction, or neutron-absorbing materials, such as boric acid in pressurized water reactors, may be introduced to the coolant system to achieve a similar effect.

Reactor technologies are classified as either thermal or fast reactors (although some technologies are "epithermal" and fall in between the two types) and by the materials used for the moderator or coolant. For example, a pressurized water reactor is a thermal reactor using water as both a coolant and moderator, and a gas-cooled fast reactor is a fast reactor using gas (carbon dioxide or helium) as a coolant.

Table 2 lists and provides information about the reactor types that are either currently operating in the United States or are advanced reactor designs under consideration for development.

Table 2. Characteristics of Commercial Reactor Types, including Those Currently Operating in the United States and Advanced Reactor Designs Under Development

Reactor type (italics indicate current commercial deployment in the United States)	Fuel	Coolant	Moderator	Process heat	Neutron spectrum
Light water reactors • *Pressurized water reactors* • *Boiling water reactors*	Uranium	Water	Water	No	Thermal
Supercritical water cooled reactor	Uranium or mixed oxide	Supercritical water	Supercritical water	Yes	Thermal (may be designed as fast)
High-temperature gas-cooled reactor / very high temperature reactor	Uranium	Helium	Graphite	Yes	Thermal
Liquid-metal-cooled fast reactor • sodium-cooled fast reactor • lead-bismuth cooled fast reactor	Uranium, plutonium, or mixed oxide	Liquid metals (sodium, lead, etc.)	N/A	No	Fast
Gas-cooled fast reactor	Uranium, plutonium, or mixed oxide	Helium or carbon dioxide	N/A	No	Fast
Molten salt reactor or molten salt breeder reactor	Uranium, thorium, or mixed oxide	Molten salts	Graphite (none for breeder)	Yes	Thermal (may be designed as fast)

Sources: GAO analysis of DOE, International Atomic Energy Agency (IAEA), and other documents. | GAO-14-545

End Notes

[1] Spent (or used) nuclear fuel—fuel that has been used in a reactor and removed from the reactor core—from light water reactors in the United States is currently stored at nuclear reactor sites around the country. About three-quarters of the fuel is stored in pools of water, and about one-quarter is stored in dry storage casks. For more information, see GAO, *Spent Nuclear Fuel: Accumulating Quantities at Commercial Reactors Present Storage and Other Challenges*, GAO-12-797 (Washington, D.C.: Aug. 15, 2012).

[2] Consistent with the Treaty on Nonproliferation of Nuclear Weapons, nuclear proliferation includes the transfer of nuclear weapons, related technology, or special fissionable material to nonnuclear weapons states. One factor contributing to the risks of nuclear proliferation and terrorism is the challenge of protecting spent nuclear fuel from diversion and theft.

[3] Light water reactors, also called Generation-II and Generation-III reactors, use ordinary (i.e., "light") water, either pressurized or boiling, for reactor moderation and cooling of the reactor core. Advanced reactors, also called Generation-IV reactors, use substances such as liquid metals or high-temperature gases to cool the reactor core. A moderator in a nuclear reactor is a substance that reduces the speed of the fast neutrons to better sustain the nuclear reaction. For more technical details of reactor types, see appendix I.

[4] For these reactors to continue operating, their owners must renew their licenses with the Nuclear Regulatory Commission (NRC), the federal agency responsible for licensing and regulating civilian nuclear reactors. In general, NRC licenses reactors to operate for 40 years and allows reactor owners to apply to renew their operating licenses for up to an additional 20 years per renewal. As of April 2014, 72 of the 100 currently operating commercial nuclear power reactors in the United States are operating with renewed operating licenses.

[5] The U.S. administration's goal is to reduce greenhouse gas emission levels by 83 percent of 2005 levels by the year 2050.

[6] GAO, Nuclear Energy: Status of DOE's Effort to Develop the Next Generation Nuclear Plant, GAO-06-1056 (Washington, D.C.: Sept. 20, 2006).

[7] The Nuclear Energy Advisory Committee includes representatives from universities, industry, foreign nationals, and national laboratories. The committee paid particular attention to obtaining a diverse membership with a balance of disciplines, interests, experiences, points of view, and geography. The committee operates in accordance with the Federal Advisory Committee Act (FACA) and all applicable FACA amendments, federal regulations, and Executive Orders.

[8] See, for example, Nuclear Energy Advisory Committee, *Report to Nuclear Energy Advisory Committee Fuel Cycle Subcommittee Meeting of April 23, 2013* (Washington, D.C.: June 13, 2013); GAO, *Nuclear Fuel Cycle Options: DOE Needs to Enhance Planning for Technology Assessment and Collaboration with Industry and Other Countries*, GAO-12-70 (Washington, D.C.: Oct. 17, 2011); GAO, *Global Nuclear Energy Partnership: DOE Should Reassess Its Approach to Designing and Building Spent Nuclear Fuel Recycling Facilities*, GAO-08-483 (Washington, D.C.: Apr. 22, 2008); *National Academies Press, Review of DOE's Nuclear Energy Research and Development Program* (Washington, D.C.: 2008).

[9] Because this was a nonprobability sample, the information gathered from these companies cannot be generalized to the nuclear power industry as a whole, but it can provide illustrative examples. Based on our review of companies developing advanced reactor technology, we selected these companies because they had involvement with DOE's advanced reactor R&D efforts.

[10] The Nuclear Energy Advisory Committee provides independent advice to the Office of Nuclear Energy on complex science and technical issues that arise in the planning, managing, and implementation of DOE's nuclear energy program. The National Research Council's mission is to improve government decision making and public policy, increase public

understanding, and promote the acquisition and dissemination of knowledge in matters involving science, engineering, technology, and health. The Next Generation Nuclear Plant Industry Alliance is an alliance of nuclear technology companies, electric utilities, and industrial end users that seek to support efforts to design, build, operate, and use high-temperature gas-cooled reactors. The Nuclear Energy Institute develops policy on legislative and regulatory issues affecting the nuclear industry and serves as the nuclear industry's voice before the U.S. Congress, executive branch agencies, and federal regulators.

[11] The DOE Office of Nuclear Energy provides funding to 10 national laboratories: (1) Argonne, (2) Brookhaven, (3) Idaho, (4) Lawrence Berkeley, (5) Lawrence Livermore, (6) Los Alamos, (7) Oak Ridge, (8) Pacific Northwest, (9) Sandia, and (10) Savannah River.

[12] The Advanced Reactor Concepts subprogram will be renamed the Advanced Reactor Technologies subprogram beginning in fiscal year 2015.

[13] Beginning in fiscal year 2015, the Advanced Small Modular Reactor subprogram will be consolidated into the Advanced Reactor Technologies subprogram.

[14] Sustainable fuel cycle options are those that would better utilize uranium resources, maximize energy generation, minimize waste generation, improve safety, and limit proliferation and terrorism risks, according to DOE.

[15] Accident-tolerant fuels are fuels that offer improved reliability and safety characteristics during accident conditions. Proliferation-resistant fuels are those that would make it more difficult or time-consuming to weaponize.

[16] In addition to these programs, NE funds programs that coordinate its international activities, manage and protect its facilities, and provide overall direction and execution of NE's programs.

[17] NE conducts most of the advanced nuclear R&D at DOE's national laboratories, but some NE-funded R&D is carried out at university laboratories, and some is done by private industry through cooperative agreements.

[18] NE's Nuclear Energy University Program was established in 2009 to consolidate university support under one initiative and better integrate university research within NE technical programs. The mission of the Nuclear Energy University Program is to engage U.S. universities in conducting R&D that supports DOE's mission, improving related infrastructure, and supporting student education.

[19] As of January 2014, the 13 participating members were Argentina, Brazil, Canada, China, Euratom (the implementing organization for development of nuclear energy within the European Union), France, Japan, Russia, South Korea, South Africa, Switzerland, the United Kingdom, and the United States, according to the Organisation for Economic Co-operation and Development (OECD).

[20] The high-temperature gas-cooled reactor is an underground helium-cooled nuclear reactor technology. The reactor and the nuclear heat supply system consist of three major components: the reactor, a heat transport system, and a cross vessel that routes the helium between the reactor and the heat transport system.

[21] Advanced reactors have been built and tested in countries around the world, including in the United States, since the 1950s. Sodium-cooled fast reactors have been built and operated for research and commercial purposes in countries around the world, including France, Japan, Germany, Russia, the United Kingdom, and the United States. Two notable U.S. fast reactors include the Experimental Breeder Reactor, operational from 1964 through 1994 and the Fast Flux Test Facility, operational from 1982 to 1992. High-temperature gas-cooled reactors have also been built and operated around the world, including two that were licensed and operated in the United States. One reactor—Peach Bottom Atomic Power Station, Unit 1—was located in Pennsylvania and operated from 1967 to 1974, according to the U.S. Nuclear Regulatory Commission. The other—Fort St. Vrain Generating Station—was located in Colorado and operated between 1979 and 1989.

[22] Pub. L. No. 109-58, §§ 641-45, 119 Stat. 594, 794-99 (2005) (codified at 42 U.S.C. §§ 16021-16025).

[23] Nuclear Energy Research Advisory Committee, *A Review of the NGNP Project* (Feb. 22, 2006).

[24] Engineering-scale demonstrations typically precede commercial-scale deployment and are meant to ensure technologies work as intended before an investment is made in a larger plant.

[25] GAO-08-483.

[26] Under the accelerated approach, DOE would partner with industry and use the latest commercially available technology to design and build a commercial-scale reprocessing plant and advanced reactor without first building engineering-scale facilities.

[27] National Research Council, *Review of DOE's Nuclear Energy Research and Development Program* (2008).

[28] EPAct 2005 also provides that DOE may submit a report to Congress establishing an alternative date for completion.

[29] The March 2012 funding opportunity announcement had a target date of 2022 for deployment. The March 2013 funding opportunity announcement had a target date of 2025 for deployment.

[30] Spent nuclear fuel can no longer efficiently generate power in a nuclear reactor, but it can be reprocessed to separate out uranium and plutonium which may be used again to fuel a reactor. Reprocessing, however, results in nuclear waste that requires disposal. The United States does not currently reprocess its spent nuclear fuel, and this fuel, when it is accepted for disposal, is considered to be high-level waste as defined by NRC.

[31] The Nuclear Waste Policy Act of 1982 directed DOE to investigate sites for a federal geologic repository to dispose of spent nuclear fuel and high-level nuclear waste from commercial nuclear power plants. In 1986, DOE recommended three candidate sites for characterization, but in 1987, the Nuclear Waste Policy Amendments Act of 1987 directed DOE to focus its efforts only on Yucca Mountain, a site about 100 miles northwest of Las Vegas, Nevada. In March 2009, the Secretary of Energy announced plans to terminate the Yucca Mountain repository program. See GAO-12-797 for more information. The administration's current strategy proposes to move spent fuel from reactor sites to an interim storage facility for eventual disposal in a geological repository underground.

[32] The International Atomic Energy Agency is an autonomous international organization affiliated with the United Nations, established in Vienna, Austria, in 1957. The agency has the dual role of promoting the peaceful uses of nuclear energy by transferring nuclear science and technology through its nuclear science and applications and technical cooperation programs, and verifying, through its safeguards program, that nuclear material subject to safeguards is not diverted to nuclear weapons or other proscribed purposes.

[33] The OECD is an organization of 34 industrialized countries, operating by consensus, that fosters dialogue among members to discuss, develop, and refine economic and social policies and provides an arena for establishing multilateral agreements.

[34] Current reprocessing techniques, aimed at extracting usable uranium and plutonium from spent fuel, have been developed for use in light water reactors. However, spent fuel from light water reactors contains other radioactive isotopes that a fast reactor can eliminate from the waste stream.

[35] As discussed above, some advanced reactor technologies have more significant technical and economic barriers to overcome, and thus are not realistic candidates for deployment in the near term. For instance, while NE could prioritize the sodium-cooled fast reactor, some industry representatives and members of the National Academy of Sciences' National Research Council told us that a fast reactor technology market does not exist in the United States. Other advanced reactors, such as fluoride salt-cooled high-temperature reactors and lead-cooled fast reactors, are not very mature or well tested, or have unproven technologies and components, according to NE officials.

[36] Isotopes are varieties of a given chemical element with the same number of protons but different numbers of neutrons. For example, the helium-3 isotope, which is used in research and to detect neutrons in radiation detection equipment, has one less neutron than the helium-4 isotope, which is the helium isotope commonly used in party balloons.

In: Advanced Reactors
Editor: Emelia Clarke

ISBN: 978-1-63463-263-8

Chapter 2

FUSION ENERGY: ACTIONS NEEDED TO FINALIZE COST AND SCHEDULE ESTIMATES FOR U.S. CONTRIBUTIONS TO AN INTERNATIONAL EXPERIMENTAL REACTOR*

United States Government Accountability Office

WHY GAO DID THIS STUDY

ITER is an international research facility being built in France to demonstrate the feasibility of fusion energy. Fusion occurs when the nuclei of two light atoms collide and fuse together at high temperatures, which results in the release of large amounts of energy. The United States has committed to providing about 9 percent of ITER's construction costs through contributions of hardware, personnel, and cash, and DOE is responsible for managing those contributions, as well as the overall U.S. fusion program. In fiscal year 2014, the U.S. ITER Project received $199.5 million, or about 40 percent of the overall U.S. fusion program budget.

GAO was asked to review DOE's cost and schedule estimates for the U.S. ITER Project. This report examines (1) how and why the estimated costs and schedule of the U.S. ITER Project have changed since 2006, (2) the reliability of DOE's current cost and schedule estimates, and (3) actions DOE has taken

* This is an edited, reformatted and augmented version of the United States Government Accountability Office publication, GAO-14-499, dated June 2014..

to reduce U.S. ITER Project costs and plan for their impact on the overall U.S. fusion program. GAO reviewed documents; assessed DOE's current estimates against best practices; and obtained the perspectives of 10 experts in fusion energy and project management.

What GAO Recommends

GAO recommends, among other things, that DOE formally propose the actions needed to set a reliable international project schedule and set a date to complete the U.S. fusion program's strategic plan. DOE agreed with GAO's recommendations.

What GAO Found

Since the International Thermonuclear Experimental Reactor (ITER) Agreement was signed in 2006, the Department of Energy's (DOE) estimated cost for the U.S. portion of ITER has grown by almost $3 billion, and its estimated completion date has slipped by 20 years. DOE has identified several reasons for the changes, such as increases in hardware cost estimates as designs and requirements have been more fully developed over time.

DOE's current cost and schedule estimates for the U.S. ITER Project reflect most characteristics of reliable estimates, but the estimates cannot be used to set a performance baseline because they are linked to factors that DOE can only partially influence. A performance baseline would commit DOE to delivering the U.S. ITER Project at a specific cost and date and provide a way to measure the project's progress. According to DOE documents and officials, the agency has been unable to finalize its cost and schedule estimates in part because the international project schedule the estimates are linked to is not reliable. DOE has taken some steps to help push for a more reliable international project schedule, such as providing position papers and suggested actions to the ITER Organization. However, DOE has not taken additional actions such as preparing formal proposals that could help resolve these issues. Unless such formal actions are taken to resolve the reliability concerns of the international project schedule, DOE will remain hampered in its efforts to create and set a performance baseline for the U.S. ITER Project.

DOE has taken several actions that have reduced U.S. ITER Project costs by about $388 million as of February 2014, but DOE has not adequately planned for the potential impact of those costs on the overall U.S. fusion program. The House and Senate Appropriations Committees have directed DOE to complete a strategic plan for the U.S. fusion program. GAO has previously reported that strategic planning is a leading practice that can help clarify priorities, and DOE has begun work on such a plan but has not committed to a specific completion date. Without a strategic plan for the U.S. fusion program, DOE does not have information to create an understanding among stakeholders about its plans for balancing the competing demands the program faces with the limited available resources or to help improve Congress' ability to weigh the trade-offs of different funding decisions for the U.S. ITER Project and overall U.S. fusion program.

Abbreviations

DOE	Department of Energy
ITER	International Thermonuclear Experimental Reactor
OMB	Office of Management and Budget
OSTP	Office of Science and Technology Policy
State	Department of State

***

June 5, 2014

The Honorable Mary Landrieu
Chair
The Honorable Lisa Murkowski
Ranking Member
Committee on Energy and Natural Resources
United States Senate

The Honorable Dianne Feinstein
Chairwoman
The Honorable Lamar Alexander
Ranking Member
Subcommittee on Energy and Water Development

Committee on Appropriations
United States Senate

The Honorable Ron Wyden
United States Senate

In 2006, the United States signed an agreement with the European Union and five other countries[1] to help build and operate the International Thermonuclear Experimental Reactor (ITER),[2] a first-of-its-kind research facility currently being built in Cadarache, France, to demonstrate the feasibility of fusion energy. Fusion occurs when the nuclei of two light atoms—typically hydrogen isotopes—collide and fuse together when heated at high temperatures and placed under tremendous pressure. The result is the release of a large amount of energy that, it is hoped, someday may be captured to produce electricity. Fusion has the potential to be an abundant source of energy to help meet future energy needs, and it also offers many other benefits. Among those benefits would be no emissions of greenhouse gases, no risk of the type of severe accidents that could occur at existing nuclear power plants, and no long-lived radioactive waste. Over the last 50 years, scientists around the world have made progress in understanding how to create the conditions for fusion, but many outstanding scientific and technical questions must still be addressed before fusion can be used as an energy source. As a result, the United States and the six other parties to the ITER Agreement identified ITER as a critical experiment that, if successful, could finally produce more power from fusion reactions than is needed to start the fusion reaction—the first step toward producing electricity from fusion energy and developing the first demonstration fusion power plant. By participating in ITER, the United States will share the cost of building this complex and expensive fusion device, benefit from the scientific and technological expertise of the six other ITER members, and have full access to ITER research results. However, since the ITER Agreement was signed in 2006, the expected total cost of building ITER has grown by billions of dollars from its original estimate of about $5 billion,[3] and its construction schedule has slipped by years, as have the cost and schedule estimates for the U.S. portion of the construction project.

The ITER Agreement established the management framework for ITER. Under the agreement, the international ITER Organization is responsible for managing the overall project, including specifying final hardware designs and performance parameters; managing the international project schedule; and assembling, installing, and commissioning the ITER facilities.[4] The ITER Organization is led by a Director-General, who serves as its Chief Executive Officer and is responsible for the execution of the organization's activities. The ITER Organization is governed by the ITER Council, which is composed of high-level government officials from each of the seven ITER members.[5] During ITER's construction phase, each of the seven ITER members is responsible for procuring and delivering an assigned set of hardware components and making cash and personnel contributions to the ITER Organization. Each member also is responsible for managing the cost and schedule of its assigned contributions within the milestones set by the international project schedule. The United States has committed to providing about 9 percent of the cost of building ITER,[6] and those construction contributions are managed by the Department of Energy (DOE) through the U.S. ITER Project Office at Oak Ridge National Laboratory in Oak Ridge, Tennessee, in partnership with Princeton Plasma Physics Laboratory and Savannah River National Laboratory.[7] DOE's Office of Fusion Energy Sciences, which is responsible for managing the overall U.S. fusion program within the Office of Science, provides program-level oversight of the U.S. ITER Project. Through March 2014, DOE had spent about $692 million on the project. In fiscal year 2014, the U.S. ITER Project received $199.5 million, which represented about 40 percent of that year's overall U.S. fusion program budget and continued a pattern of substantial funding increases for the project that began in fiscal year 2011 after several years of fluctuating funding levels.

In 2007, when the U.S. ITER Project was just beginning, we reported on the importance of DOE assessing the full costs of U.S. participation in ITER and setting a definitive cost estimate for the project.[8] We noted at the time that DOE had made a commitment to provide hardware components to ITER without a definitive cost and schedule estimate and a complete project design and, as a result, DOE's preliminary cost estimate of $1.122 billion for U.S. contributions to ITER's construction might be subject to significant change. We also noted the risk that management challenges facing the ITER Organization could result in ITER construction delays and further increase costs for the United States. Today, significant questions remain about how much the U.S. ITER Project will cost, when it will be completed, and how DOE plans to manage the impact of U.S. ITER Project costs on the overall U.S. fusion program in a constrained federal budget environment. The House and Senate Appropriations Committees, in particular, have expressed concern that DOE will fund the U.S. ITER Project at the expense of the rest of the U.S.

fusion program, and they have directed DOE to provide detailed cost and schedule estimates for the project.

In this context, you asked us to review DOE's cost and schedule estimates for U.S. contributions to ITER's construction. This report examines (1) how and why the estimated cost and schedule for the U.S. ITER Project have changed since 2006; (2) the reliability of DOE's current cost and schedule estimates for the U.S. ITER Project and the factors, if any, that have affected their reliability; and (3) the actions DOE has taken, if any, to reduce U.S. ITER Project costs and plan for their potential impact on the overall U.S. fusion program.

To conduct this work, we reviewed the ITER Agreement, relevant laws, and DOE guidance. To determine how and why the estimated costs and schedule for the U.S. ITER Project have changed since 2006, we reviewed DOE and U.S. ITER Project Office documents. We assessed the reliability of the data on the changes in the cost estimates by interviewing DOE officials and U.S. ITER Project Office representatives who had knowledge of the data and by other means, and we determined that the data were sufficiently reliable for reporting on the reasons for the changes in the cost estimates. To evaluate the reliability of DOE's current cost and schedule estimates and the factors, if any, that have affected their reliability, we reviewed DOE's most recent cost and schedule estimates for the U.S. ITER Project—as developed by the U.S. ITER Project Office in August 2013—and DOE's internal peer review of those estimates. We also reviewed DOE's project management order and related guidance, as well as an October 2013 report to the ITER Council on the results of a management assessment of the ITER Organization. We assessed the reliability of DOE's current cost and schedule estimates by analyzing those estimates against best practices identified in GAO's *Cost Estimating and Assessment Guide and Schedule Assessment Guide*.[9] Specifically, we reviewed documentation, interviewed the U.S. ITER Project Office representatives who prepared the estimates, reviewed relevant sources, and compared the information collected with the best practices. For DOE's current schedule estimate, we assessed the characteristics of two selected subordinate schedules—the central solenoid modules schedule and the tokamak cooling water system schedule—that are used as inputs to the integrated U.S. ITER Project master schedule.[10] We were able to use the results of the two subordinate schedules to provide insight into the integrated master schedule since the same strengths and weaknesses of the subordinate schedules would transfer to the master schedule. We determined that the schedules were sufficiently reliable for our reporting purposes, and our report notes the instances where reliability concerns affect the quality of the schedules. To examine the actions DOE has taken, if any, to reduce U.S. ITER Project costs and plan for their potential impact on the overall U.S. fusion program, we reviewed DOE and U.S. ITER Project Office documents on actions taken to reduce and plan for U.S. ITER Project costs and reviewed our prior work on leading practices in federal strategic planning for agency divisions, programs, or initiatives.[11] We also interviewed 10 experts in fusion energy and the management of large scientific research projects. We selected these experts through a multistep process to ensure coverage and a range of perspectives from industry, DOE's national laboratories, and universities. We analyzed the experts' responses across a standard set of questions and summarized the results. Appendix I presents a more detailed description of our objectives, scope, and methodology, and appendix II lists the names and affiliations of the 10 experts we interviewed.

We conducted this performance audit from June 2013 to June 2014 in accordance with generally accepted government auditing standards. Those standards require that we plan and perform the audit to obtain sufficient, appropriate evidence to provide a reasonable basis for our findings and conclusions based on our audit objectives. We believe that the evidence obtained provides a reasonable basis for our findings and conclusions based on our audit objectives.

BACKGROUND

Fusion is the energy source that powers the sun and stars. Fusion occurs when the nuclei of two light atoms collide with sufficient energy to overcome their natural repulsive forces and fuse together. Scientists are currently using deuterium and tritium—two hydrogen isotopes—for this reaction. When the nuclei of the two atoms collide, the collision produces helium and a large quantity of energy (see figure 1). For the fusion reaction to take place, the atoms must be heated to very high temperatures—about 100 million degrees centigrade, or 10 times the temperature of the surface of the sun—and placed under tremendous pressure.

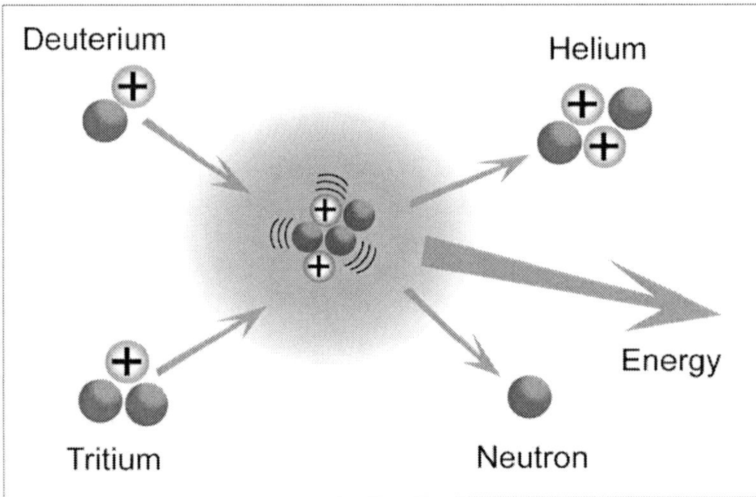

Source: Copyright © ITER Organization. All rights reserved. | GAO-14-499

Figure 1. The Fusion Reaction.

For more than 50 years, the United States has been trying to control fusion to produce electricity. The United States is pursuing two paths to achieve controlled fusion—magnetic and inertial. The goal for both approaches is to generate more energy than is needed to begin and sustain the fusion reaction. The world's first controlled release of fusion power was achieved in 1991, but no fusion device has succeeded in generating more power than it consumes. Magnetic fusion uses magnetic devices to confine a plasma, consisting of electrically charged atoms, and sustain a fusion reaction.[12] ITER will be a magnetic fusion device known as a "tokamak."[13] To reduce the risk of investing in only one device, DOE's Office of Fusion Energy Sciences also funds scientific research on alternative types of magnetic devices. Inertial fusion relies on intense lasers or particle beams to heat and compress a small, frozen pellet of deuterium and tritium—a few millimeters in size—that would yield a burst of energy. The lasers or particle beams would continuously heat and compress the pellets, which would simulate, on a very small scale, the actions of a hydrogen bomb. The National Nuclear Security Administration, a separately organized agency within DOE, is leading efforts in inertial fusion because it can be used for defense needs, such as validating the integrity and reliability of the U.S. nuclear weapons stockpile.

ITER is considered to be the next step in magnetic fusion. It is an experiment to study fusion reactions in conditions similar to those expected in a future electricity-generating power plant. The goal is to be the first fusion device in the world to produce a substantial amount of net power—that is, produce more power than it consumes. Specifically, the objective is to produce 10 times more power than is needed to start the fusion reaction in pulses of 5 or more minutes. ITER also will test a number of key technologies, including the heating, control, and remote maintenance systems that will be needed for a fusion power station. ITER has been planned to consist of four phases: (1) construction, (2) operation, (3) deactivation, and (4) decommissioning. The construction phase, which is the sole focus of the U.S. ITER Project, began in 2007. The international project schedule, as of April 2014, anticipates that the ITER fusion device will be built by 2019 and achieve its "first plasma" in 2020.[14] After ITER has achieved its first plasma, the next several years are expected to be devoted to a preliminary period of operation in pure hydrogen during which physics testing will be done, followed by operation in deuterium with a small amount of tritium to test ITER's wall shielding. This will then be followed by the start of full ITER operations in an equal mixture of deuterium and tritium, at which point ITER will be used to try to produce 10 times more power than it consumes. As of April 2014, the international project schedule anticipates the start of deuterium-tritium operations in 2027. ITER's operation phase is expected to last 20 years followed by a 5-year deactivation phase and then a decommissioning phase. If ITER is successful, it will lead to power plant design and testing.

ITER was first proposed at the Geneva Summit in November 1985, when President Reagan and Soviet Premier Gorbachev recognized that joint activities were needed to diffuse the tension of the arms race during the cold war and begin the then Soviet Union's economic integration into the world economy. The goal was to share scientific and technical information in a program where both sides had a comparable level of knowledge and could jointly realize commercial gains from the development of fusion technology. Following the Geneva Summit, the United States, the Soviet Union, Japan, and several European countries entered into an agreement to develop ITER's design. The United States stopped participating in the project in 1999 after Congress raised concerns that the technical basis for the proposed project was not sound, the cost was too high, and the facility was too large. In response to the U.S. withdrawal, the countries participating in ITER reduced the size of the facility and the estimated cost of building ITER to about $5 billion. Subsequently, a number of scientific advances increased U.S. confidence that the new ITER design would meet its scientific and technological goals and, in January 2003, President Bush announced that the United States would rejoin

ITER. In 2003, the People's Republic of China and the Republic of Korea also joined the project; in December 2005, India became the seventh and most recent member to join. In November 2006, all six countries and the European Union signed the ITER Agreement. Figure 2 shows the countries participating in ITER and the location of the ITER site.

Under the ITER Agreement, the United States is responsible for 9.09 percent of ITER's total construction cost.[15] DOE fulfills this obligation by supplying personnel to work for the ITER Organization; by making cash contributions to the ITER Organization to cover common expenses; and by procuring and delivering the 12 assigned U.S. hardware components to the ITER Organization (see table 1 for a list of the 12 U.S. hardware components).[16] The ITER Agreement includes a provision that allows any ITER member except the European Union to withdraw from the project after the agreement has been in force for 10 years, which would be in October 2017.[17] However, withdrawing members still have responsibility for providing the entire cost of their assigned hardware components and cash contributions to the construction phase, and they could be responsible for other costs as well if they withdraw during ITER's operation phase.

Table 1. U.S. Hardware Components to Be Delivered to the ITER Organization

U.S. hardware component	Purpose
Central solenoid magnet system	Confines, shapes, and controls the plasma inside ITER's vacuum vessel.
Diagnostics (15% of ITER diagnostics)	Provides the measurements necessary to control, evaluate, and optimize plasma performance and to further the understanding of plasma physics.
Disruption mitigation systems (up to a capped value)	Limits the impact of plasma current disruptions to the tokamak vacuum vessel and other components.
Electron cyclotron heating transmission lines	Brings additional heat to the plasma and deposits heat in specific areas of the plasma to minimize instabilities.
Ion cyclotron heating transmission lines	Brings additional heat to the plasma.
Pellet injection system	Provides efficient fueling and delivers hydrogen, deuterium, or a deuterium/tritium mixture as required by plasma operations.

U.S. hardware component	Purpose
Roughing pump system	Exhausts certain parts of the ITER machine.
Steady state electrical network (75% of total network)	Supplies the electricity needed to operate the entire ITER plant, including offices and the operational facilities.
Tokamak cooling water system	Manages temperatures generated during the operation of the tokamak.
Tokamak exhaust processing system	Separates tokamak exhaust gases into a stream containing only hydrogen isotopes and a stream containing only nonhydrogen gases.
Toroidal field conductor (8% of total conductor)	Part of the toroidal field magnet that confines, shapes, and controls the plasma inside the ITER vacuum vessel.
Vacuum auxiliary systems	Creates low density in ITER's vacuum vessel and connected vacuum components.

Source: GAO analysis of DOE documents. / GAO-14-499

The percentage of the overall U.S. fusion program budget going toward the U.S. ITER Project has grown since fiscal year 2011 after fluctuating in previous years (see figure 3). In fiscal year 2011, the U.S. ITER Project received \$80 million, or about 22 percent, of the U.S. fusion program budget of roughly \$367 million. In fiscal year 2012, the U.S. ITER Project received \$105 million, or about 27 percent, of the U.S. fusion program budget of approximately \$393 million. In fiscal year 2013, the project received about \$124 million, or roughly 33 percent, of the U.S. fusion program budget of approximately \$378 million and, in fiscal year 2014, it received just under \$200 million, or about 40 percent, of the U.S. fusion program budget of approximately \$505 million. DOE has requested that \$150 million, or about 36 percent, of its fiscal year 2015 proposed budget for the U.S. fusion program (\$416 million) go toward the U.S. ITER Project.

DOE's project management order[18] establishes, among other things, a process for planning and executing large construction projects.[19] The order establishes several major milestones—or "critical decision points"—that span the life of the project and includes DOE's review and approval of a project's cost and schedule estimates at the major milestones. The order emphasizes that cost estimates should be provided at each major milestone, but the degree of rigor and uncertainty of the estimates will depend on the stage of the project. Specifically, it provides that cost and schedule estimates are not considered to be final until DOE has approved a "performance baseline" for a project.[20] A performance baseline captures a project's key performance, scope, cost, and schedule parameters, and it represents a commitment from DOE to Congress to deliver a project within those parameters. DOE's project management order explains that a performance baseline sets the bar against which a project's progress will be measured, and that cost and schedule estimates developed prior to DOE approval of a performance baseline are only preliminary estimates. According to DOE guidance,[21] developing a performance baseline for a project

- fosters effective coordination and integration of planning and decision making among the different DOE parties involved in a project;
- enables routine monitoring and reporting on all aspects of a project's performance;
- helps define a project and ensure that project requirements are rigorously refined; and
- facilitates the delivery of first-of-a-kind projects that typically have a high degree of uncertainty, high cost, high impact on stakeholders, and high visibility.

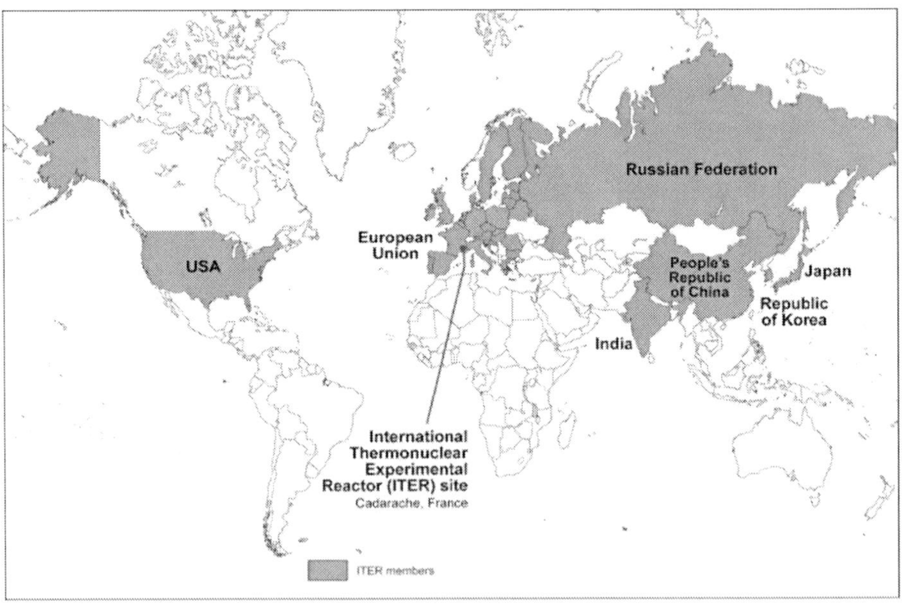

Source: GAO ; Map Resources (map). | GAO-14-499

Figure 2. Countries Participating in ITER and ITER Site Location.

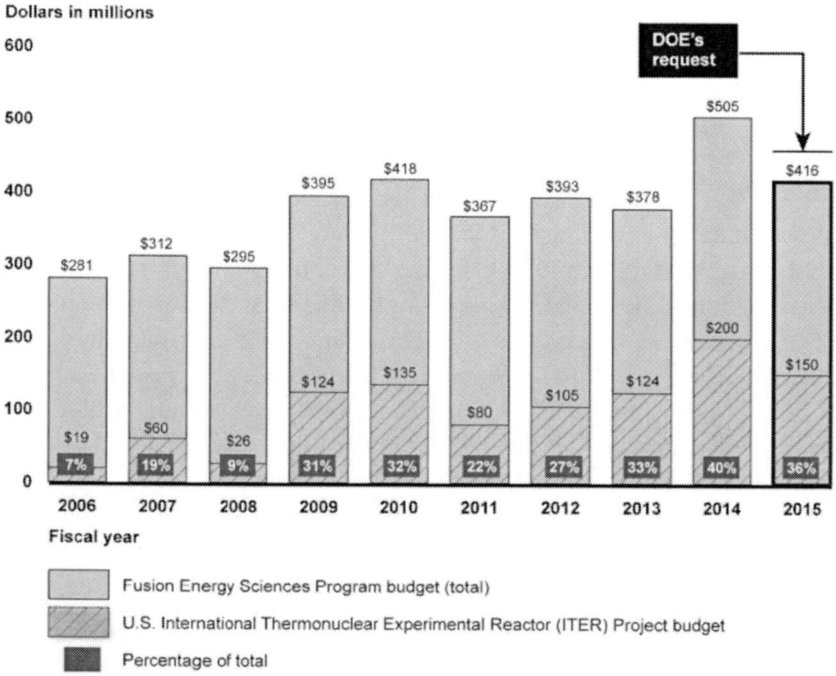

Source: GAO analysis of DOE data. | GAO-14-499

Note: This figure reports dollar values rounded to millions of dollars and percentages rounded to the nearest whole percentage point.

Figure 3. U.S. ITER Project Share of the U.S. Fusion Program Budget, Fiscal Years 2006-2015.

THE ESTIMATED COST AND SCHEDULE OF THE U.S. ITER PROJECT HAS GROWN SUBSTANTIALLY SINCE 2006

The estimated cost and schedule of the U.S. ITER Project has grown substantially since the ITER Agreement was signed in 2006 (see figure 4), and DOE has identified several reasons for these changes. At the time the ITER Agreement was signed, DOE planned on spending $1.122 billion on the U.S. ITER Project and expected to complete the project in 2013[22] based on preliminary estimates it approved in 2005.[23] In 2008, DOE formally increased its preliminary cost estimate to a cost range of from $1.45 billion to $2.2 billion.[24] This was the most recent time DOE approved a cost estimate for the

U.S. ITER Project.[25] Also, in 2008, DOE said it expected ITER to achieve first plasma in 2016 and expected the U.S. ITER Project to be completed in 2017. DOE now estimates that the U.S. ITER Project will cost $3.915 billion, that ITER will achieve first plasma in 2023, and that the U.S. ITER Project will be completed in 2033. These estimates were developed based on a set of key assumptions that were not used by DOE in its 2008 estimates. The current estimates include an assumption of annual funding for the U.S. ITER Project of $225 million, starting in fiscal year 2015 and continuing until the project is completed, with no adjustment for inflation; limited U.S. cash contributions to the ITER Organization from 2014 to 2016; future U.S. contributions to the ITER Organization for ITER's operation, deactivation, and decommissioning phases coming out of the $225 million annual funding level; and a hardware delivery schedule not tied to the ITER Organization's schedule. DOE also instructed the U.S. ITER Project Office to use its estimate of ITER achieving first plasma in 2023 rather than the 2020 date in the ITER Organization's schedule, and to use its best estimate for ITER Organization construction costs rather than the budget most recently approved by the ITER Council.[26] Nonetheless, these current estimates remain preliminary, according to DOE officials, because DOE has not approved a performance baseline for the U.S. ITER Project. A performance baseline captures a project's key performance, scope, cost, and schedule parameters, and represents a commitment from DOE to Congress to deliver a project within those parameters.

According to DOE documents, the current $3.915 billion cost estimate for the U.S. ITER Project includes the following:

- $1.469 billion (38 percent)[27] to complete the remaining work to procure and deliver U.S. hardware components for ITER;
- $928 million (24 percent) in contingency to address potential schedule delays or increases in costs for manufacturing components, including $852 million in contingency for the remaining work to procure and deliver U.S. hardware components, and $76 million in contingency for U.S. cash contributions to the ITER Organization;[28]
- $541 million (14 percent) to account for project costs through June 2013;
- $519 million (13 percent) for remaining cash contributions to the ITER Organization to pay for scientists, engineers, and support personnel working at the ITER Organization; the assembly and installation of the components in France to build the reactor; quality

assurance testing of all ITER member-supplied components; and contingencies; and

- $458 million (12 percent) for escalation costs, such as changes in currency exchange rates and commodity prices, which are driven by the extended length of the project due to the funding assumptions used to develop the cost estimate.

DOE documents and officials identified several key reasons for the growth of the cost and schedule estimates for the U.S. ITER Project. DOE officials identified a number of reasons that led to the change in the project's initial 2005 cost estimate of $1.122 billion to its 2008 cost range of from $1.45 billion to $2.2 billion. These included (1) higher estimates for the cost of U.S. hardware components as their designs and requirements were more fully developed; (2) updated estimates for external factors such as currency exchange rates and commodity prices; (3) changes to U.S. hardware component requirements and the international project schedule due to a 2007 design review of the overall ITER project; and (4) additions for contingency and recognized risks. According to DOE documents and officials, the primary reasons the U.S. ITER Project's cost estimate grew from the upper range of $2.2 billion in 2008 to the current $3.915 billion estimate included the following:[29]

- *Higher estimates for U.S. hardware components as designs and requirements have been more fully developed over time*: The design and requirements for U.S. hardware components have evolved over time, and the cost estimates for these components have changed as a result, according to DOE documents and officials. Specifically, the current U.S. ITER Project cost estimate includes about $770 million more than the 2008 figures due to greater understanding of what U.S. hardware components are likely to cost. According to DOE officials, expected hardware costs now reflect more fully developed designs, better industry estimates of the cost of producing those designs, and greater understanding of project risks, among other things. DOE documents noted that, as of February 2014, about two-thirds of U.S. hardware components by value were in final design or beyond, although there were some smaller components that were still in earlier stages of design.[30] According to DOE officials, because of the progress in developing hardware component designs, the project team has a greater understanding of what those components are likely to cost and has reflected that understanding in the current $3.915 billion cost estimate.

- *Higher contingency amounts added to address risks*: The U.S. ITER Project Office has increased the amount of contingency in the current cost estimate to address the risks from the project's significantly longer schedule and to increase confidence in the estimate, according to DOE documents and officials. Specifically, the current estimate includes about $681 million more in contingency than was in the U.S. ITER Project cost estimate in 2008. According to DOE officials, the amount of contingency is based on detailed risk analyses as well as a management assessment of key assumptions. DOE officials told us that, to develop contingency amounts for the current estimate, the U.S. ITER Project Office identified and evaluated program-level risks, such as changes in currency exchange rates and growth in ITER Organization cash contribution requirements. They also considered project-level risks, such as the U.S. ITER Project's dependence on other ITER members or the ITER Organization for inputs needed to complete U.S. hardware components and potential procurement and manufacturing difficulties. Further, they added contingency amounts if U.S. ITER Project Office management representatives thought initial contingency amounts were too low for the risks associated with the project's longer schedule or based on their past experience with large science projects.
- *Schedule delays:* U.S. schedule delays due to international project schedule delays and U.S. funding constraints accounted for about $544 million of the increase in the U.S. ITER Project's cost estimate since DOE last approved the estimate in 2008, according to DOE documents and officials. First, international project schedule delays have lengthened the U.S. ITER Project schedule and have also led to increases in the cost estimate. For example, in 2007, an international review identified extensive changes that were needed in ITER's design, and that significantly delayed the ITER Organization from defining requirements for U.S. hardware components.[31] This in turn created delays in the U.S. ITER Project schedule for procuring and delivering those components, which led to higher cost estimates for those components. Second, U.S. funding constraints resulting from the project's most recent $225 million per year funding plan and lower-than-requested funding levels in some years have lengthened the U.S. ITER Project schedule, according to DOE documents and

officials. This in turn has made it necessary for the U.S. ITER Project Office to build additional amounts into the project's cost estimate to account for higher escalation costs and the longer period of time the U.S. ITER Project workforce will be needed. For example, DOE officials told us that the project's most recent $225 million per year funding plan reflected discussions between DOE, the Department of State (State), the Office of Management and Budget (OMB), and the Office of Science and Technology Policy (OSTP) to provide enough funding to meet U.S. obligations to ITER and reduce the amount of the U.S. fusion program budget and the overall DOE budget that had to be devoted to the U.S. ITER Project on an annual basis.[32] However, this lengthened the U.S. ITER Project's schedule and created procurement inefficiencies, resulting in increases in the project's overall cost estimate.

- *Higher cash contributions to the ITER Organization due to growth in ITER construction costs*: The U.S. ITER Project Office built an additional $348 million into the current cost estimate to reflect the increase in U.S. cash contributions it expects to have to make to the ITER Organization, according to DOE documents and officials. DOE officials explained this increase includes $169 million for the U.S. share of a previously approved increase in the ITER Organization's construction budget, as well as $179 million for the U.S. share of a potential future billion euro increase in the ITER Organization construction budget and the anticipated cost increases for the ITER Organization staff based on a 2023 first plasma date.

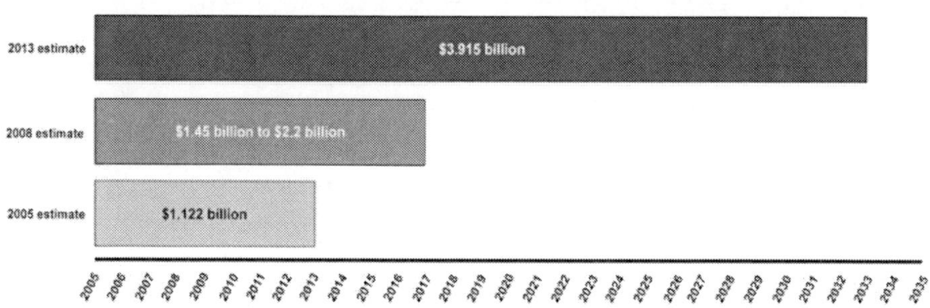

Source: GAO analysis of DEO data. | GAO-14-499

Figure 4. Total Estimated Cost and Completion Date for the U.S. ITER Project.

Table 2. Four Characteristics of High-Quality, Reliable Cost and Schedule Estimates

Characteristics of reliable cost estimates	
Comprehensive	A comprehensive cost estimate has enough detail to ensure that cost elements are neither omitted nor double counted.
Well-documented	A well-documented cost estimate allows for data it contains to be traced to source
	documents.
Accurate	An accurate cost estimate is based on an assessment of most likely costs and has been adjusted properly for inflation.
Credible	A credible cost estimate discusses any limitations because of uncertainty or bias surrounding data or assumptions.
Characteristics of reliable schedule estimates	
Comprehensive	A comprehensive schedule includes all government and contractor activities necessary to accomplish a project's objectives.
Well-constructed	A well-constructed schedule sequences all activities using the most straightforward logic possible.
Credible	A credible schedule uses data about risks and opportunities to predict a level of confidence in meeting the completion date.
Controlled	A controlled schedule is updated periodically to realistically forecast dates for activities.

Source: GAO. / GAO-14-499

- *The United States has taken on additional hardware responsibilities*: DOE agreed to take on additional hardware responsibilities, accounting for $39 million of the increase in the U.S. ITER Project cost estimates since 2008, according to DOE documents and officials.[33] DOE officials told us that there is a cost cap for each of these additional hardware responsibilities, and that the ITER Organization will be responsible for paying any amounts that exceed the caps. As a result, they view taking on these additional hardware responsibilities as a way to reduce uncertainty about future U.S. ITER Project costs because DOE will not have to spend more than the amounts specified in the cost caps for each item.

Despite Reflecting Most Characteristics of Reliable Cost and Schedule Estimates, DOE's Estimates Cannot Be Used to Set a Baseline in Part Because They Are Linked to an Unreliable International Schedule

DOE's current cost estimate for the U.S. ITER Project reflects most of the characteristics of a reliable cost estimate, and its schedule estimates reflect all characteristics of a reliable schedule.[34] However, DOE's estimates cannot be used to set a performance baseline that would commit DOE to delivering the project at a specific cost and date primarily because of some factors that DOE can only partially influence. The factors DOE can only partially influence include an unreliable international project schedule to which the U.S. schedule is linked and an uncertain U.S. funding plan. DOE has taken some action to address the factors that have prevented it from setting a performance baseline and finalizing its estimates, but significant challenges remain.

DOE's Current Cost Estimate for the U.S. ITER Project Reflects Most Characteristics of a Reliable Estimate, and Its Schedule Estimates Reflect All These Characteristics

DOE's current cost estimate for the U.S. ITER Project reflects most of the characteristics of high-quality, reliable cost estimates as established by best practices documented in the *GAO Cost Estimating and Assessment Guide*.[35] In addition, DOE's current schedule estimates fully reflect the characteristics of high-quality, reliable schedule estimates as established by best practices documented in the *GAO Schedule Assessment Guide*.[36] According to the guides, four characteristics make up reliable cost estimates—they are comprehensive, well-documented, accurate, and credible (see table 2). Similarly, four characteristics make up reliable schedule estimates—they are comprehensive, well-constructed, credible, and controlled. Cost and schedule estimates are considered reliable if each of the four characteristics is substantially or fully met. If any of the characteristics is not met, minimally met, or partially met, then the estimates do not fully reflect the characteristics of a high-quality estimate and cannot be considered reliable.

DOE's current cost estimate for the U.S. ITER Project—as developed by the U.S. ITER Project Office in August 2013—substantially met best practices for comprehensive, well-documented, and accurate estimates, but only partially met best practices for credible estimates. For example, DOE's cost estimate substantially met best practices for documenting all assumptions that will influence costs (comprehensive), describing step by step how the estimate was developed (well-documented), and adjusting properly for inflation (accurate). However, the U.S. ITER Project Office did not conduct a complete sensitivity analysis on the cost estimate, and an independent cost estimate has not been conducted (credible).[37] The U.S. ITER Project Office did identify four key assumptions from the estimate for sensitivity testing. However, the analysis did not include some cost elements that represent high percentages of the overall estimate, including some of the most expensive hardware components being built by the United States. For example, the sensitivity analysis did not include the tokamak cooling water system, which is the most expensive U.S. hardware component. Without a comprehensive sensitivity analysis that identifies how the cost estimate is affected by changes to its assumptions, DOE will not fully understand how certain risks can affect the cost estimate and potentially result in decisions based on incomplete information. In addition, DOE did not conduct an independent cost estimate to determine whether other estimating methods produce similar results. DOE policy does not require an independent cost estimate until it approves a performance baseline, which the agency does not expect to occur until late 2015. However, including an independent estimate is a best practice associated with credible cost estimates. Independent cost estimates are less likely to reflect organizational bias. They also incorporate adequate risk, which generally results in more conservative estimates due to higher estimated costs. Without such an independent cost estimate, DOE faces a greater risk of underfunding the project, which can lead to overall cost growth and schedule slippage. (See app. IV for the individual ratings of each cost estimating practice.)

DOE's current schedule estimates for the two most expensive U.S. hardware items[38]—the central solenoid modules and the tokamak cooling water system—fully met best practices for comprehensive schedules and substantially met best practices for well-constructed, credible, and controlled schedules. For example, DOE's schedule estimate fully met best practices for capturing and establishing the duration of all activities (comprehensive), and substantially met best practices for sequencing all activities (well-constructed), conducting a schedule risk analysis (credible), and updating the schedule with actual progress (controlled). However, the schedule estimate partially met best practices for horizontal and vertical traceability,[39] maintaining a baseline schedule, and ensuring reasonable total float, which is the amount of time an activity can be delayed before the dates of the program's completion milestones are affected. U.S. ITER Project Office representatives acknowledged these issues with the schedule and attributed them to problems with the international project schedule. For example, according to project representatives, DOE's schedule does not align with the international project schedule (i.e.: is not vertically traceable) because the international project schedule does not account for delays in ITER Organization delivery milestones, including a 30-month delay in ITER construction site preparations. Without up-to-date reliable international milestones, DOE cannot develop realistic U.S. milestones that align with the international project schedule and set a baseline that can provide a reliable, specific cost and completion date for the project. (See app. V for the individual ratings of each scheduling best practice).

DOE's Cost and Schedule Estimates Cannot Be Used to Set a Baseline Due to Factors That DOE Can Only Partially Influence

DOE considers its current cost and schedule estimates for the U.S. ITER Project to be preliminary, and these estimates cannot be used to set a performance baseline that would represent a commitment from DOE to Congress to deliver the project at a specific cost and date. DOE policy says that cost and schedule estimates are considered final only after a performance baseline has been approved for a project, and DOE has not approved a performance baseline for the U.S. ITER Project. According to DOE's project management order, a performance baseline sets a bar against which a project's progress can be measured. DOE's target date for setting a performance baseline finalizing its cost and schedule estimates for the U.S. ITER Project has continually slipped from an original expected date of fiscal year 2007 to the current target of late in fiscal year 2015. That is when DOE expects the international project schedule to be updated, according to DOE documents and officials.

According to DOE documents and officials, DOE's current estimates for the U.S. ITER Project cannot be used to set a performance baseline because of three factors, two of which DOE can only partially influence as follows:

- First, the overall international project schedule that DOE uses as a basis for the U.S. schedule is not reliable. In July 2010, the ITER

Council approved an official schedule for the overall ITER project. However, an October 2013 management assessment of the ITER Organization determined that the international project schedule established in 2010 was not reliable. The assessment attributed the unreliable schedule, in part, to management deficiencies within the ITER Organization. For example, the assessment found that the ITER Organization's senior management had insisted that the international project schedule not be changed even when staff had developed what they thought were more realistic schedules, and that staff had not been allowed to openly challenge the schedule. According to DOE officials, the ITER Organization plans to spend the next year reassessing the international project schedule and taking actions to address the identified management deficiencies, and it hopes to complete its schedule reassessment by June 2015.[40]

- Second, DOE has not proposed a final, stable funding plan for the U.S. ITER Project. DOE's most recent plan had been to provide a flat $225 million per year for the project, and that figure was the basis for its current cost and schedule estimates. However, DOE officials told us that this funding plan could potentially change depending on the outcome of the ITER Organization's reassessment of the international project schedule. In March 2014, DOE requested $150 million for the U.S. ITER Project in fiscal year 2015, $75 million less than the $225 million per year funding plan. According to DOE documents, the $150 million request would allow the U.S. ITER Project to meet its fiscal year 2015 commitments to ITER but would not be enough for the project to meet the milestones set in the official international project schedule. Officials noted that if Congress provided less than DOE's requested funding, the U.S. ITER Project schedule will slip further. Six of the 10 fusion energy and project management experts we interviewed said that identifying sufficient funding to execute the U.S. ITER Project poses a significant management challenge for DOE.

The third factor that has kept DOE from setting a performance baseline finalizing its estimates is within the agency's direct control. Specifically, an August 2013 internal peer review found that the methodologies used to develop DOE's current cost and schedule estimates were appropriate, but that the estimates do not sufficiently consider all project risks and uncertainties.[41] For example, the review found that the U.S. ITER Project Office did not identify and quantify all risks, that its view of potential risk mitigation was too optimistic, and that the range of possible cost outcomes due to each individual risk factor was too narrow. Further, the review identified potential cost increases related to changing technical requirements, uncertainty about the ITER Organization's performance, and the dependence on other ITER members for production of items that are used in U.S. hardware components. To better account for these risks and uncertainties, the review added additional amounts to DOE's current cost estimate of $3.915 billion and found that the U.S. ITER Project was more likely to cost from $4 billion to $6.5 billion.[42] The reviewers recommended, among other things, that the U.S. ITER Project Office update its risk estimates to be more comprehensive and reevaluate its risk mitigation assessments before DOE approves a performance baseline for the U.S. ITER Project.

In the absence of a performance baseline, DOE has developed a 2-year plan for the U.S. ITER Project that sets near-term cost and schedule targets to guide the project's performance in fiscal years 2013 and 2014.[43] However, this 2-year plan is an interim measure and does not represent DOE's commitment to a specific cost and schedule for the U.S. ITER Project as a performance baseline would. Most of the fusion energy and project management experts we interviewed emphasized the importance of DOE approving a performance baseline for the U.S. ITER Project, with some experts noting that a performance baseline would provide a goal for all project stakeholders to work toward and might ease concerns about the uncertainty of the funding levels needed to complete the project. Several experts also told us that, until DOE approves a performance baseline for the U.S. ITER Project, there will continue to be uncertainty about the project's direction.

DOE Has Taken Some Actions to Address the Factors Preventing It from Setting a Baseline, but Challenges Remain

DOE has taken some actions to address the factors preventing it from setting a performance baseline that would allow the agency to finalize its cost and schedule estimates for the U.S. ITER Project. However, project management and schedule deficiencies in the ITER Organization and uncertainty in the U.S. ITER Project funding plan continue to pose management challenges for the agency and delay its efforts to set a performance baseline.

According to DOE officials, DOE has taken several actions to try to get the ITER Organization to address international project management and scheduling deficiencies. For example, DOE officials told us that their aggressive participation in early ITER Agreement negotiations led to the adoption of a biannual management assessment requirement. This has focused attention on international management deficiencies, resulting in several recommendations for improving the project. Further, DOE officials told us they have used ITER Council Management Advisory Committee meetings to introduce, communicate, and advance project management principles, such as competitive procurement actions, in an effort to improve ITER Organization project management.[44] Additionally, DOE has developed position papers describing the agency's concerns with ineffective ITER Organization scheduling and management and suggesting actions the ITER Organization could take to develop a reliable international project schedule and improve international management. For example, in a position paper on scheduling issues, DOE recommended the ITER Council direct the ITER Organization to focus on developing a short-term schedule, and defer long-term schedule development until lessons are learned from the short-term effort. DOE has provided the position papers to other ITER members and achieved some unofficial support, but the agency has not submitted a formal proposal on the suggested actions to the ITER Council, which could vote on and ultimately require the implementation of these actions. According to DOE officials, DOE has not submitted formal proposals because previous ITER Council Chairs delayed substantive discussions of issues such as schedule slippages and conducted meetings with a primary goal of obtaining consensus among ITER members. However, DOE officials said the ITER Council and other ITER members are aware of DOE's position, and they hope the new ITER Council Chair, who took over in January 2014,[45] will change the way the ITER Council operates. DOE officials also said the ITER Council approved at the November 2013 ITER Council meeting the initiative for the ITER Organization to develop a short-term annual work plan for 2014, the results of which will inform long-term schedule development, and that all milestones had been met by all seven ITER members for the first three months.

Even so, challenges remain that will continue to hamper DOE's ability to develop a baseline for the U.S. ITER Project. Eight out of the 10 fusion energy and project management experts we interviewed said DOE does not have enough information from the ITER Organization and other ITER members to effectively plan the U.S. ITER Project, and 7 of the 10 experts said the international structure and management issues contribute to DOE's management challenges. In this context, DOE's efforts may have helped improve ITER Organization project management and helped jump-start efforts to develop a reliable international project schedule, but a reliable international project schedule is not expected until June 2015, as previously noted. Further, the previous international project schedule developed by the ITER Organization and approved by the ITER Council in 2010 has not proved reliable, and management issues that were identified in previous years continue to pose challenges at the international level. For example, the October 2013 management assessment of the ITER Organization found that the ITER Council had not acted on many recommendations for project management improvements from a previous management assessment in 2009 and that the problems identified in that assessment continue. The most recent management assessment attributed the inaction to the ITER Council's reliance on consensus decision making, which caused it to avoid or delay difficult decisions. The 2013 assessment stated that the ITER members needed to openly discuss and then make decisions on difficult issues at ITER Council meetings, even if there is no consensus, and all members should be held accountable for results. The management assessment contained 11 recommendations that were designed to be taken together. The assessment further said that the international project would not achieve significant improvement if the ITER Council only adopted a few recommendations, as has been the case with recommendations in previous management assessments. DOE officials told us that the ITER Council had approved proposals to address the recommendations from the October 2013 management assessment and that the ITER Council Chair was actively monitoring the implementation of the recommendations.

To address the uncertainty of the U.S. funding plan for the U.S. ITER Project, DOE has evaluated a range of funding scenarios for executing the project. As previously noted, DOE most recently developed a $225 million per year flat funding plan with State, OMB, and OSTP, but DOE officials acknowledged that the plan constrained funding for the project to allow DOE to meet U.S. obligations to ITER and reduce the amount of the U.S. fusion program and overall DOE budgets devoted to the project annually. DOE officials told us the flat funding plan created a long and inefficient schedule and has created gaps between the design and fabrication stages of some systems, which has led to cost growth. DOE officials said that they will not be able to meet the most recent international project schedule under this funding plan, and delivery of some U.S. components will likely be late under the plan. DOE officials told us they plan to develop a final, stable funding plan for the

U.S. ITER Project, but that plan can only be developed if the international project schedule is reliable.

To ensure that all risks and uncertainties are sufficiently incorporated into its estimates, DOE officials told us the U.S. ITER Project Office held a series of risk workshops. According to DOE officials, as of March 2014, the U.S. ITER Project Office had held risk workshops on U.S. hardware components and associated risks, as well as a workshop on external risks. U.S. ITER Project Office representatives told us that they are currently analyzing the results of the workshops and that the workshops will ultimately lead to updated risk estimates and an update to the current cost estimate of $3.915 billion for the U.S. ITER Project.

DOE Has Taken Actions to Reduce U.S. ITER Project Costs but Has Not Adequately Planned for Their Impact on the U.S. Fusion Program

DOE has taken several actions to reduce the cost of the U.S. ITER Project. Some fusion energy and project management experts we interviewed suggested additional strategies DOE could pursue that might further reduce U.S. ITER Project costs or improve project management. DOE has not adequately planned for the potential impact of U.S. ITER Project costs on the overall U.S. fusion program because it has not completed a strategic plan that would clarify the program's priorities given those costs.

DOE Has Taken Several Actions to Reduce U.S. ITER Project Costs

According to DOE documents and officials, DOE has taken several actions to reduce the cost of the U.S. ITER Project by about $388 million as of February 2014,[46] including the following:

- *Value engineering*: The U.S. ITER Project Office has identified ways to design U.S. hardware components that lower costs but maintain the component's essential functions. According to DOE officials, this strategy—known as "value engineering"—has been an integral part of the U.S. design effort.[47] For example, the U.S. ITER Project Office has been able to reduce the cost of the central solenoid magnet system by more than $18 million by eliminating, simplifying, or reducing the number of some parts. It also has reduced the cost of the vacuum auxiliary systems by almost $34 million by, among other things, revising test equipment items and quantities and reducing the number of system connections and pumps. According to DOE officials, these and other value engineering efforts have resulted in about $225 million in savings as of February 2014. DOE documents indicated that about half of the savings come from value engineering the design of the tokamak cooling water system.

- *Centralized and consolidated procurement for certain items*: The U.S. ITER Project Office has agreed to have the ITER Organization centrally procure piping for the tokamak cooling water and vacuum systems for which the United States is responsible. It has also reached agreement with other ITER members to consolidate procurement of certain common parts rather than having each ITER member procure those parts for their assigned hardware components. For example, the U.S. ITER Project Office has agreed to have the European Union procure all cable trays needed for U.S. hardware components. According to DOE officials, centralized and consolidated procurement for certain items has saved the U.S. ITER Project about $120 million as of February 2014.
- *Scope transfers and reallocations*: The U.S. ITER Project Office has reached agreements to transfer some U.S. hardware responsibilities to the ITER Organization and to reallocate some hardware responsibilities among ITER members to improve procurement efficiency and reduce U.S. costs. For example, project officials told us that they reached an agreement to shift the U.S. responsibility for procurement of one system to other ITER members, with the United States taking on more engineering work for that system in return. DOE officials estimated that this resulted in $20 million in savings. According to DOE officials, scope transfers and reallocations have saved about $43 million as of February 2014.
- *Other strategies*: According to DOE documents and officials, the U.S. ITER Project Office has used several other strategies to reduce U.S. ITER Project costs. These strategies have included implementing lessons learned and leading practices from other large projects; working with the ITER Organization on cost reduction initiatives to improve ITER Organization processes and requirements; and

minimizing costs associated with project execution by, for example, providing incentives in contract provisions. DOE officials said that it was too soon to quantify the cost savings resulting from these actions.

According to DOE documents and officials, these cost savings are reflected in the current $3.915 billion cost estimate for the U.S. ITER Project. DOE officials told us that cost containment will continue to be a high priority for the project. They told us that the most significant opportunity to further reduce U.S. ITER Project costs would be the adoption of an optimal funding plan for the project. Officials explained that $458 million of the current $3.915 billion cost estimate is included to cover escalation costs, and an optimal funding plan could potentially reduce those and other costs by allowing the U.S. ITER Project to be completed in a shorter period of time.

Some Experts Suggested Strategies to Further Reduce U.S. ITER Project Costs or Improve Management

Some of the 10 fusion energy and project management experts that we interviewed identified strategies DOE could pursue to further reduce the cost of the U.S. ITER Project. The following two strategies were suggested by several of these experts:

- Six experts suggested that DOE could reduce U.S. ITER Project costs by adopting an optimal funding plan for the project. The optimal funding plan being suggested would involve different dollar figures year to year rather than DOE's most recent strategy of funding the U.S. ITER Project at a flat $225 million per year starting in fiscal year 2015. Some experts noted that a funding plan that scaled up in the near-term and then scaled down in later years could potentially reduce overall U.S. ITER Project costs by hundreds of millions of dollars. Specifically, three experts said that an optimal funding plan could shorten the current U.S. ITER Project schedule, and that doing so would reduce overall project costs. Two experts noted that one issue with this strategy would be that the U.S. ITER Project would need to receive more funding in some years, and that could lead to funding cuts for other DOE projects and programs. To address that issue, one expert suggested DOE not propose an optimal funding plan for the U.S. ITER Project until the agency determines that there is a reliable international project schedule and that the ITER Organization has made significant progress improving its management of the overall ITER project. Another expert suggested that DOE first communicate its plans for managing the impact on the overall U.S. fusion program of higher U.S. ITER Project funding in certain years before proposing an optimal funding plan. When we discussed this strategy with DOE program officials, they agreed that an optimal funding plan for the project could shorten the U.S. ITER Project schedule and potentially reduce costs significantly. However, they emphasized that the most recent $225 million annual funding plan was an attempt to balance the funding needs of the U.S. ITER Project and the rest of the U.S. fusion program.

- Three experts suggested that DOE could reduce U.S. ITER Project costs by working with its international partners to develop a reliable international project schedule and aligning U.S. ITER Project efforts with that schedule. One expert noted that this would help make DOE's cost estimate for the U.S. ITER Project more certain. Another expert told us that a reliable international project schedule was necessary for DOE to set an optimal funding plan, which in turn could help minimize U.S. ITER Project costs. The expert said that aligning the U.S. ITER Project with a reliable international project schedule was necessary to ensure that the United States was not producing its hardware components too early or too late, either of which could result in cost growth. When we discussed this strategy with DOE program officials, they said that the United States has actively presented its views to its international partners on what would be a realistic schedule for the overall ITER project. Further, the officials told us that they do try to align U.S. ITER Project efforts with the international project schedule in spite of imperfect knowledge due to the current international project schedule not being reliable.

Some of the 10 experts we interviewed also suggested strategies that DOE could pursue to improve its management of the U.S. ITER Project. The following two strategies were mentioned by several experts:

- Six experts suggested DOE establish a separate office that would report directly to top DOE management officials to provide oversight of the U.S. ITER Project. Two of these experts said that having a separate office would give the U.S. ITER Project greater visibility at the highest levels of DOE. Another expert said a benefit would be to enhance the project's interaction with stakeholders, including Congress and the U.S. fusion community. Further, two experts told us that a separate office for the U.S. ITER Project would enhance DOE's ability to oversee the project's complexity and complications given the international structure the project operates within. When we discussed this strategy with DOE program officials, they said that a separate DOE office to oversee the U.S. ITER Project would not provide many benefits and could have unintended consequences. Specifically, they said that the project already has high visibility with top DOE management officials and that creating a separate office could result in a greater degree of funding competition between the U.S. ITER Project and the rest of the U.S. fusion program.

- Four of the 10 experts we interviewed said DOE should make more information on the U.S. ITER Project available to stakeholders in the U.S. fusion community. One expert said DOE should be more forthcoming about what the agency expects the U.S. ITER Project to cost and how they plan to pay for it. Two other experts suggested that DOE should disclose its internal peer reviews of the U.S. ITER Project. Several of the experts we interviewed also identified a number of negative effects of DOE not being sufficiently transparent about the U.S. ITER Project. These included the erosion of stakeholder commitment to the project and a diminished ability for stakeholders to effectively plan research efforts and make informed funding decisions related to the impact of project costs on the overall U.S. fusion program. When we discussed this strategy with DOE program officials, they said they have shared a substantial amount of information on the U.S. ITER Project with stakeholders and were trying to share more. For example, they noted that DOE included a detailed section on the U.S. ITER Project in its fiscal year 2015 budget request. In some cases, however, officials said they were limited in the information they could share with stakeholders by the international sensitivities of the overall ITER project, by procurement sensitivities, and by the nature of the budget process.

DOE Has Not Adequately Planned for the Impacts of U.S. ITER Project Costs on the U.S. Fusion Program

DOE has not adequately planned for the potential impact of U.S. ITER Project costs on the overall U.S. fusion program, although it has taken some steps. For example, as previously mentioned, the project's most recent $225 million per year funding plan was an attempt by DOE and other parts of the administration to meet U.S. obligations to ITER and reduce the project's impact on the U.S. fusion program and DOE's overall budget. However, according to agency officials, DOE has not completed a strategic plan for the U.S. fusion program to clarify the program's goals and priorities and its proposed approach for meeting them in light of the potential impact of U.S. ITER Project costs.[48]

The House and Senate Appropriations Committees directed DOE to submit a 10-year strategic plan for the U.S. fusion program in the explanatory statements that accompanied both the fiscal year 2012 and fiscal year 2014 energy and water development appropriation acts.[49] In addition, DOE's Fusion Energy Sciences Advisory Committee recommended in 2013 that DOE develop a strategic plan for the U.S. fusion program using the advisory committee process and with broad U.S. fusion community input.[50] Further, 9 of the 10 fusion energy and project management experts we interviewed agreed that it would be useful for DOE to develop such a plan.

DOE officials told us they have not completed a strategic plan for the U.S. fusion program to date, for three reasons. First, they said an effort in 2012 to obtain the Fusion Energy Sciences Advisory Committee's input on U.S. fusion program priorities had been unsuccessful because the committee did not address program priorities under a constrained budget scenario due to conflict of interest issues. Second, DOE officials said there had been too much budget uncertainty in fiscal year 2013 regarding the U.S. ITER Project and the overall U.S. fusion program to complete a plan. They explained that the House and Senate Appropriations Committees proposed different direction and funding levels for the U.S. fusion program in fiscal year 2013, and these differences were not resolved until the passage of the fiscal year 2014 appropriations act in January 2014. Third, DOE officials said an effort in 2012 and 2013 to develop a high-level strategic document for the U.S. fusion program was unsuccessful because OMB did not concur with the document developed by DOE. DOE officials said they are in the early stages of developing a strategic plan in response to the House and Senate appropriations committees' direction in the explanatory statement that accompanied the fiscal year 2014 appropriations act. These officials said they would ask the Fusion Energy Sciences Advisory Committee in April 2014 to provide input on U.S. fusion program priorities by October 2014. According to the officials, DOE will consider the committee's input in developing a strategic plan for the U.S. fusion program, and it hopes to finalize a plan no later than the January 2015 deadline set by the House and Senate Appropriations Committees. However, DOE officials could not provide a specific date when they expect to complete the strategic plan.

We have previously reported that strategic planning is a leading management practice organizations can employ to define their mission and goals; help clarify priorities; address management and other challenges that threaten an agency's ability to meet its long-term strategic goals; align activities, core processes, and resources to accomplish those goals; and foster informed communication between an agency and its stakeholders.[51] Without a strategic plan for the overall U.S. fusion program that addresses DOE's plans for managing the impacts of U.S. ITER Project costs, the agency does not have information to

- involve and help create a basic understanding among stakeholders—including Congress and the U.S. fusion community—about its plans for balancing the competing demands that confront the program with the limited resources available;
- better ensure that the U.S. ITER Project and other U.S. fusion program activities are aligned to effectively and efficiently achieve the program's goals; and
- improve Congress's ability to weigh the potential trade-offs of different funding decisions for the U.S. ITER Project and the overall U.S. fusion program within a constrained budget environment.

CONCLUSION

DOE has taken some actions to address the factors that affect the reliability of its cost and schedule estimates for the U.S. ITER Project. However, more than 7 years and nearly $700 million after the ITER Agreement was signed, significant uncertainty remains about how much the U.S. ITER Project will cost, when it will be completed, and how DOE plans to manage the impact of the project's costs on the overall U.S. fusion program. DOE's current preliminary cost and schedule estimates met most characteristics of high-quality, reliable cost and schedule estimates, but the cost estimate was not fully credible. Specifically, the U.S. ITER Project Office did not include the most expensive U.S. hardware component in its sensitivity analysis and an independent cost estimate has not been conducted, which could result in DOE making decisions based on incomplete information and increase the risk of more cost growth.

It is important for DOE to set a performance baseline for the U.S. ITER Project in order to finalize its cost and schedule estimates, provide a bar against which the project's progress can be measured, and allow Congress to make well-informed funding decisions about the project within a constrained budget environment. However, DOE has not yet set a performance baseline for the U.S. ITER Project in part because the international project schedule is not reliable, a key factor that DOE can only partially influence. To its credit, DOE has taken several actions to push for a reliable international project schedule and improvements to ITER Organization project management. Nonetheless, the agency could do more to ensure the ITER Organization develops a reliable international project schedule and that ITER Organization project management deficiencies are addressed by making formal proposals to the ITER Council that address these issues and remaining vigilant about the timely implementation of the proposed improvements. Without a reliable international project schedule, DOE neither can propose a final, stable funding plan for the U.S. ITER Project, nor can it reasonably assure Congress that the project's cost will not continue to grow and the schedule will not continue to slip.

DOE has taken some steps to reduce the cost of the U.S. ITER Project and plan for the impact of the project's cost on the overall U.S. fusion program. However, even though there has been repeated direction from the House and Senate Appropriations Committees going back more than 2 years and a recommendation from its own advisory committee to do so, DOE has not yet completed a strategic plan for the overall U.S. fusion program. Strategic planning is a leading practice that can help organizations clarify priorities and address challenges that threaten their ability to meet long-term strategic goals. Completing a strategic plan for the overall U.S. fusion program would reduce uncertainty by addressing DOE's priorities for the program in light of U.S. ITER Project costs. Moreover, involving stakeholders, such as the Fusion Energy Sciences Advisory Committee, in the plan's development would increase stakeholder understanding of DOE's plans for balancing the competing demands that face the U.S. fusion program with the limited resources available. DOE is beginning the initial work on such a plan, but a similar effort that was started in 2012 did not result in a completed strategic plan for the U.S. fusion program, and the agency has not provided a specific date when it will complete its current effort. Without committing to a specific date, DOE may not complete a strategic plan for the U.S. fusion program in a timely manner and, without a completed strategic plan, DOE may face challenges ensuring that it has effectively aligned U.S. fusion program activities to achieve program goals. Further, Congress and the U.S. fusion community are likely to remain uncertain about DOE's plans for balancing the competing funding demands of the U.S. ITER Project and the rest of the U.S. fusion program.

RECOMMENDATIONS FOR EXECUTIVE ACTION

To reduce uncertainty about the expected cost and schedule of the U.S. ITER Project and its potential impact on the U.S. fusion program, the Secretary of Energy should direct the Associate Director of the Office of Fusion Energy Sciences to take the following four actions:

- Direct the U.S. ITER Project Office to revise and update the project's cost estimate to meet all characteristics of high-quality, reliable cost estimates. Specifically, the U.S. ITER Project Office should revise the project's cost estimate to ensure it is credible by including a comprehensive sensitivity analysis that includes all significant cost elements and conducting an independent cost estimate;
- Develop and present at the next ITER Council meeting a formal proposal describing the actions DOE believes need to be taken to set a reliable international project schedule and improve ITER Organization project management. Continue to formally advocate for the timely implementation of those actions at each future ITER Council meeting until the ITER Council approves an updated international project schedule;
- Once the ITER Organization completes its reassessment of the international project schedule, use that schedule, if reliable, to propose a final, stable funding plan for the U.S. ITER Project, approve a performance baseline with finalized cost and schedule estimates, and communicate this information to Congress; and
- Set a specific date for completing, in a timely manner, a strategic plan for the U.S. fusion program that addresses DOE's priorities for the

overall U.S. fusion program in light of U.S. ITER Project costs, and involve the Fusion Energy Sciences Advisory Committee in the development of the plan.

AGENCY COMMENTS AND OUR EVALUATION

We provided a draft copy of this report to DOE for review and comment. DOE provided written comments on the draft report on May 27, 2014 and also provided technical and clarifying comments, which we incorporated as appropriate. DOE agreed with each of the report's recommendations and said it has taken steps or plans to take additional steps to fully implement them.

Frank Rusco
Director, Natural Resources and Environment

APPENDIX I: OBJECTIVES, SCOPE, AND METHODOLOGY

Our review assessed: (1) how and why the estimated cost and schedule for the U.S. International Thermonuclear Experimental Reactor (ITER) Project have changed since 2006; (2) the reliability of the Department of Energy's (DOE) current cost and schedule estimates for the U.S. ITER Project and the factors, if any, that have affected their reliability; and (3) the actions DOE has taken, if any, to reduce U.S. ITER Project costs and plan for their potential impact on the overall U.S. fusion program. To address these objectives and better understand the U.S. ITER Project, we reviewed the ITER Agreement, relevant laws, and DOE guidance and met with DOE and Department of State officials, representatives from the U.S. ITER Project Office, and fusion energy and project management experts from industry, DOE's national laboratories, and universities.

To determine how and why the estimated cost and schedule for the U.S. ITER Project have changed since 2006, we reviewed DOE and U.S. ITER Project Office documents. We assessed the reliability of the data on changes in the cost estimates by checking for obvious errors in accuracy and completeness; comparing the data with other sources of information; and interviewing DOE officials and U.S. ITER Project Office representatives who had knowledge of the data. We determined that DOE and the U.S. ITER

Project Office's data on changes in the cost estimates for the U.S. ITER Project were sufficiently reliable for reporting on the reasons for the changes in the estimates. We also contacted the national audit offices of each of the six other ITER members to identify any audit reports they had issued on ITER, and we reviewed each report that was identified.

To evaluate the reliability of DOE's current cost and schedule estimates and the factors, if any, that have affected their reliability, we reviewed DOE's most recent cost and schedule estimates for the U.S. ITER Project—as developed by the U.S. ITER Project Office in August 2013—and DOE's internal peer review of those estimates. We also reviewed DOE's project management order and related guidance, as well as an October 2013 report to the ITER Council on the results of a management assessment of the ITER Organization. We assessed the reliability of DOE's current cost and schedule estimates by analyzing the August 2013 estimates against the best practices identified in GAO's *Cost Estimating and Assessment* Guide (Cost Guide) and *Schedule Assessment Guide (Schedule Guide).*[52]

Specifically, we determined the reliability of the cost estimate by reviewing documentation DOE submitted for the cost estimate, interviewing U.S. ITER Project Office representatives who prepared the estimate, reviewing relevant sources, and comparing the information collected with the best practices identified in the Cost Guide to determine whether the cost estimate was (1) comprehensive, (2) accurate, (3) well-documented, and (4) credible.[53] After a review of all source data, we assessed the extent to which the cost estimate met these best practices by calculating the assessment rating of each criteria within the four characteristics on a five-point scale: not met = 1, minimally met = 2, partially met = 3, substantially met = 4, and fully met = 5.[54] Then, we took the average of the individual assessment ratings for the criteria to determine the overall rating for each of the four characteristics. The resulting average became the overall assessment as follows: not met = 0 to 1.4; minimally met = 1.5 to 2.4; partially met = 2.5 to 3.4; substantially met = 3.5 to 4.4; and fully met = 4.5 to 5.0. After conducting our initial analysis, we shared it with DOE officials and representatives from the U.S. ITER Project Office who developed the cost estimate to provide an opportunity for them to comment and identify reasons for observed shortfalls in cost estimating best practices. We took their comments and any additional information they provided and incorporated it into the assessments to finalize the scores for each characteristic and best practice.

We determined the reliability of the current schedule estimate for the U.S. ITER Project by assessing the characteristics of two selected subordinate schedules—the central solenoid modules schedule and the tokamak cooling water system schedule—that are used as inputs to the integrated master schedule. We selected these two schedules because they are the largest two hardware items, in terms of value, that the United States is responsible for contributing to ITER. We determined whether the schedules were (1) comprehensive, (2) well-constructed, (3) credible, and (4) controlled by reviewing documentation DOE submitted for the U.S. ITER Project schedule estimate, interviewing U.S. ITER Project Office representatives who developed the estimate, reviewing relevant sources, and comparing the information collected against the criteria for each of these characteristics identified in the *Schedule Guide*.[55] We also analyzed schedule metrics as a part of that analysis to highlight potential areas of strengths and weaknesses against each of our four characteristics of a reliable schedule. In order to assess each schedule against the four characteristics and their accompanying 10 best practices, we traced and verified underlying support and determined whether the U.S. ITER Project Office provided sufficient evidence to satisfy the criterion and assigned a score based on the same five-point scale we used in our analysis of the cost estimate. Then, we took the average of the individual assessment ratings to determine the overall rating for each of the characteristics, also using the same scale we used in our analysis of the cost estimate. After conducting our initial analysis, we shared it with DOE officials and representatives from the U.S. ITER Project Office who developed the schedule estimate to provide an opportunity for them to comment and identify reasons for observed shortfalls in schedule management best practices. We took their comments and any additional information they provided and incorporated it into the assessments to finalize the scores for each characteristic and best practice. By examining the two subordinate schedules against our guidance, we conducted a reliability assessment on each of the schedules and incorporated our findings on reliability limitations in the analysis of each subordinate schedule. We were also able to use the results of the two subordinate schedules to provide insight into the health of the integrated master schedule since the same strengths and weaknesses of the subordinate schedules would transfer to the master schedule. We determined that the schedules were sufficiently reliable for our reporting purposes, and our report notes the instances where reliability concerns affect the quality of the schedules.

To examine the actions DOE has taken, if any, to reduce U.S. ITER Project costs and plan for their potential impact on the overall U.S. fusion program, we reviewed DOE and U.S. ITER Project Office documents on actions taken to reduce U.S. ITER Project costs, interviewed DOE program officials about the status of their efforts to complete a strategic plan for the U.S. fusion program, and reviewed meeting records of DOE's Fusion Energy Sciences Advisory Committee.[56] We also reviewed our prior work on leading practices in federal strategic planning for agency divisions, programs, or initiatives,[57] as well as the House and Senate Appropriations Committees' direction to DOE in the explanatory statements for the fiscal year 2012 and fiscal year 2014 energy and water development appropriation acts to submit a 10-year strategic plan for the U.S. fusion program. Further, we summarized the results of semistructured interviews with 10 experts in fusion energy and the management of large scientific research projects. To select these experts, we first identified 105 experts by reviewing the results of a literature search; congressional hearings; National Academies of Science publications; membership lists for the Fusion Energy Sciences Advisory Committee, the ITER Council Management Advisory Committee,[58] the ITER Council Science and Technology Advisory Committee,[59] and DOE's internal peer reviews of the U.S. ITER Project; and recommendations from other fusion energy experts we interviewed. From this list, we then used a multistep process to select 10 experts.[60] To ensure coverage and a range of perspectives, we selected fusion energy and large scientific project management experts from industry, DOE's national laboratories, and universities. We conducted semistructured interviews with the 10 selected experts using a standard set of questions and analyzed their responses, grouping them into overall themes. We summarized the results of our analysis and then asked DOE program officials for their views on actions suggested by multiple stakeholders to potentially reduce U.S. ITER Project costs or improve U.S. ITER Project management. Not all 10 of the experts answered all of our questions. The views expressed by experts do not represent the views of GAO. Appendix II lists the names and affiliations of the 10 experts we interviewed.

We conducted this performance audit from June 2013 to June 2014 in accordance with generally accepted government auditing standards. Those standards require that we plan and perform the audit to obtain sufficient, appropriate evidence to provide a reasonable basis for our findings and conclusions based on our audit objectives. We believe that the evidence obtained provides a reasonable basis for our findings and conclusions based on our audit objectives.

APPENDIX II: LIST OF EXPERTS

- Dr. Riccardo Betti, University of Rochester
- Dr. Aesook Byon, Brookhaven National Laboratory (retired)
- Dr. Richard W. Callis, General Atomics
- Dr. Adam Cohen, Princeton Plasma Physics Laboratory
- Dr. Ray Fonck, University of Wisconsin—Madison
- Dr. Charles M. Greenfield, General Atomics
- Dr. Martin Greenwald, Massachusetts Institute of Technology
- Dr. Richard Hawryluk, Princeton Plasma Physics Laboratory
- Dr. Robert Iotti, CH2M Hill (retired)
- Dr. Dale Meade, Princeton Plasma Physics Laboratory (retired)

Appendix III: Status of U.S. Hardware Components for the International Thermonuclear Experimental Reactor (ITER) as of August 2013

Dollars in millions

U.S. hardware component	Purpose	Design status	Estimate to complete[a]	Estimated final delivery date	Procurement arrangement[b] reached
Central solenoid magnet system	Confines, shapes, and controls the plasma inside ITER's vacuum vessel.	Final design[c, f]	$252.1	April 2019	Yes
Diagnostics (15% of ITER diagnostics)	Provides the measurements necessary to control, evaluate, and optimize plasma performance and to further the understanding of plasma physics.	Preliminary design[d]	$112.2	September 2024	Yes
Disruption mitigation systems (up to a capped value)	Limits the effect of plasma current disruptions to the tokamak vacuum vessel and other components.	Preliminary design	$19.4	June 2019	No
Electron cyclotron heating transmission lines	Brings additional heat to the plasma and deposits heat in specific areas of the plasma to minimize instabilities.	Final design	$90.4	November 2024	Yes
Ion cyclotron heating transmission lines	Brings additional heat to the plasma.	Preliminary design	$98.1	August 2026	Yes

Dollars in millions

U.S. hardware component	Purpose	Design status	Estimate to complete[a]	Estimated final delivery date	Procurement arrangement[b] reached
Pellet injection system	Provides efficient fueling and delivers hydrogen, deuterium, or a deuterium/tritium mixture as required by plasma operations.	Preliminary design	$39.0	July 2026	No
Roughing pump system	Exhausts certain parts of the ITER machine.	Conceptual design[e]	$36.0	September 2021	No
Steady state electrical network (75% of total network)	Supplies the electricity needed to operate the entire ITER plant, including offices and the operational facilities.	Design complete[f]	$39.2	March 2016	Yes
Tokamak cooling water system	Manages temperatures generated during the operation of the tokamak.	Final design[f]	$463.4	September 2029	Yes
Tokamak exhaust processing system	Separates tokamak exhaust gases into a stream containing only hydrogen isotopes and a stream containing only non-hydrogen gases.	Preliminary design	$107.0	April 2030	No

Dollars in millions					
U.S. hardware component	**Purpose**	**Design status**	**Estimate to complete[a]**	**Estimated final delivery date**	**Procurement arrangement[b] reached**
Toroidal field conductor (8% of total conductor)	Part of toroidal field magnet that confines, shapes, and controls the plasma inside the ITER vacuum vessel.	Design complete[f]	$10.3	July 2015	Yes
Vacuum auxiliary systems	Creates low density in ITER's vacuum vessel and connected vacuum components.	Preliminary design[f]	$58.7	September 2023	Yes

Source: GAO analysis of DOE documents.

[a] The "Estimate to complete" refers to the latest estimate of budget required to finish the work remaining. The estimates listed in this column are burdened, meaning they include direct and indirect costs, and unescalated, meaning they represent current year dollar values.

[b] "Procurement arrangements" govern the procurement of plant systems, components, or site construction, detailing all the necessary technical specifications and management requirements.

[c] "Final design" is the last stage of design development prior to implementation. At the final design stage, the project scope should be finalized and changes should be permitted only for compelling reasons.

[d] The "Preliminary design" stage initiates the process of converting concepts to a more detailed design whereby more detailed and reliable cost and schedule estimates are developed. This stage of the design is complete when it provides sufficient information to support development of the performance baseline.

[e] The "Conceptual design" process must ensure that a solution or alternatives are responsive to an approved need, and also technically achievable, affordable, and will provide the best value to the agency.

[f] Certain parts of these components are already being fabricated or being prepared for fabrication.

APPENDIX IV: GAO ASSESSMENT OF DOE'S CURRENT (AUGUST 2013) COST ESTIMATE FOR THE U.S. INTERNATIONAL THERMONUCLEAR EXPERIMENTAL REACTOR (ITER) PROJECT COMPARED WITH BEST PRACTICES

Characteristic	Overall assessment[a]	Best practice	Individual assessment
Comprehensive	*Substantially met*	The cost estimate includes all life-cycle costs.	*Partially met*
		The cost estimate completely defines the program, reflects the current schedule, and is technically reasonable.	*Substantially met*
		The cost estimate work breakdown structure is product-oriented, traceable to the statement of work/objective, and at an appropriate level of detail to ensure that cost elements are neither omitted nor double-counted.	*Substantially met*
		The estimate documents all cost-influencing ground rules and assumptions.	*Substantially met*
Well-documented	*Substantially met*	The documentation should capture the source data used, the reliability of the data, and how the data were normalized.	*Substantially met*
		The documentation describes in sufficient detail the calculations performed and the estimating methodology used to derive each element's cost.	*Substantially met*
		The documentation describes step by step how the estimate was developed so that a cost analyst unfamiliar with the program could understand what was done and replicate it.	*Substantially met*
		The documentation discusses the technical baseline description and the data in the baseline is consistent with the estimate.	*Fully met*
		The documentation provides evidence that the cost estimate was reviewed and accepted by management.	*Substantially met*
Accurate	*Substantially met*	The cost estimate results are unbiased, not overly conservative or optimistic and based on an assessment of most likely costs.	*Partially met*

Characteristic	Overall assessment[a]	Best practice	Individual assessment
		The estimate has been adjusted properly for inflation.	*Substantially met*
		The estimate contains few, if any, minor mistakes.	*Fully met*
		The cost estimate is regularly updated to reflect significant changes in the program so that it is always reflecting current status.	*Substantially met*
		Variances between planned and actual costs are documented, explained, and reviewed.	*Substantially met*
		The estimate is based on a historical record of cost estimating and actual experiences from other comparable programs.	*Substantially met*
		The estimating technique for each cost element was used appropriately.	*Partially met*
Credible	*Partially met*	The cost estimate includes a sensitivity analysis that identifies a range of possible costs based on varying major assumptions, parameters, and data inputs.	*Partially met*
		A risk and uncertainty analysis was conducted that quantified the imperfectly understood risks and identified the effects of changing key cost driver assumptions and factors.	*Partially met*
		Major cost elements were cross-checked to see whether results were similar.	*Partially met*
		An independent cost estimate was conducted by a group outside the acquiring organization to determine whether other estimating methods produce similar results.	*Partially met*

Source: GAO analysis of DOE data.

[a] Not met– DOE provided no evidence that satisfies any of the criterion; minimally met– DOE provided evidence that satisfies a small portion of the criterion; partially met– DOE provided evidence that satisfies about half of the criterion; substantially met– DOE provided evidence that satisfies a large portion of the criterion; and fully met– DOE provided complete evidence that satisfies the entire criterion.

APPENDIX V: GAO ASSESSMENT OF DOE'S CURRENT (AUGUST 2013) SCHEDULE ESTIMATES FOR THE TWO LARGEST HARDWARE ITEMS OF THE U.S. INTERNATIONAL THERMONUCLEAR EXPERIMENTAL REACTOR (ITER) PROJECT COMPARED WITH BEST PRACTICES

Characteristic	Overall assessment[a]		Best practice	Individual assessment	
	Central solenoid modules	**Tokamak cooling water system**		**Central solenoid modules**	**Tokamak cooling water system**
Comprehensive	*Fully met*	*Fully met*	Capturing all activities.	*Fully met*	*Fully met*
			Assigning resources to all activities.	*Substantially met*	*Substantially met*
			Establishing the durations of all activities.	*Fully met*	*Fully met*
Well-constructed	*Substantially met*	*Substantially met*	Sequencing all activities.	*Substantially met*	*Substantially met*
			Confirming that the critical path is valid.	*Substantially met*	*Substantially met*
			Ensuring reasonable total float.	*Partially met*	*Partially met*
Credible	*Substantially met*	*Substantially met*	Verifying that the schedule is traceable horizontally and vertically.	*Partially met*	*Partially met*
			Conducting a schedule risk analysis.	*Substantially met*	*Substantially met*
Controlled	*Substantially met*	*Substantially met*	Updating the schedule with actual progress and logic.	*Substantially met*	*Fully met*
			Maintaining a baseline schedule.	*Partially met*	*Partially met*

Source: GAO analysis of DOE data.

[a] Not met– DOE provided no evidence that satisfies any of the criterion; minimally met– DOE provided evidence that satisfies a small portion of the criterion; partially met– DOE provided evidence that satisfies about half of the criterion; substantially met – DOE provided evidence that satisfies a large portion of the criterion; and fully met– DOE provided complete evidence that satisfies the entire criterion.

End Notes

[1] The five countries are India, Japan, the People's Republic of China, the Republic of Korea, and the Russian Federation.

[2] The term "International Thermonuclear Experimental Reactor" is no longer used for the project. Instead, the project is simply referred to as "ITER."

[3] According to a Department of Energy (DOE) document, the original estimate of about $5 billion for ITER's construction was in 2002 dollars and based on the 2001 ITER design. The estimate was developed for the purpose of establishing relative contributions of the ITER members toward ITER's construction and was not comparable to a DOE construction project cost estimate.

[4] The ITER Organization will also be responsible for managing the operation and deactivation of the ITER facilities.

[5] The ITER Council meets at least twice a year. ITER members have the ability to raise issues for consideration and decision by the council.

[6] Of the seven ITER members, the United States and five other countries—the People's Republic of China, Japan, India, the Republic of Korea, and the Russian Federation—are each providing 9.09 percent of the total construction cost. The European Union is the largest contributor—45.46 percent—because ITER is being built on a European Union country's soil, and the European Union has agreed to pay for infrastructure costs.

[7] The U.S. ITER Project covers only U.S. contributions to ITER's construction phase. U.S. contributions for ITER's operation, deactivation, and decommissioning phases are not part of the U.S. ITER Project.

[8] GAO, *Fusion Energy: Definitive Cost Estimates for U.S. Contributions to an International Experimental Reactor and Better Coordinated DOE Research Are Needed,* GAO-08-30 (Washington, D.C.: Oct. 26, 2007).

[9] GAO, *GAO Cost Estimating and Assessment Guide: Best Practices for Developing and Managing Capital Program Costs,* GAO-09-3SP (Washington, D.C.: Mar. 2, 2009) and *GAO Schedule Assessment Guide: Best Practices for Project Schedules—Exposure Draft,* GAO-12-120G (Washington, D.C.: May 30, 2012).

[10] We selected these two subordinate schedules because they are the largest two hardware items, in terms of value, that the United States is responsible for contributing to ITER. By examining the two subordinate schedules against our guidance, we conducted a reliability assessment on each of the schedules and incorporated our findings on reliability limitations in the analysis of each subordinate schedule.

[11] For example, see GAO, *Environmental Protection: EPA Should Develop a Strategic Plan for Its New Compliance Initiative*, GAO-13-115 (Washington, D.C.: Dec. 10, 2012) and *Environmental Justice: EPA Needs to Take Additional Actions to Help Ensure Effective Implementation*, GAO-12-77 (Washington, D.C.: Oct. 6, 2011).

[12] Current magnetic devices have not been able to sustain this fusion reaction for more than a few seconds, not the long period of time that the reaction would need to be sustained to produce electricity.

[13] The term "tokamak" comes from a Russian acronym for a fusion device that was developed in the former Soviet Union during the 1950s and 1960s. A tokamak has been the most successful magnetic fusion device, but there is still uncertainty that it will lead to a commercially viable fusion energy device.

[14] "First plasma" refers to the first time the ITER device is able to successfully produce a plasma. Construction will continue after this to prepare ITER to use deuterium-tritium fuel. The introduction of deuterium-tritium fuel marks the start of full ITER operations.

[15] The United States will also be responsible in the future for 13 percent of the costs of operating, deactivating, and decommissioning ITER. These costs—which DOE estimated would be about $1.5 billion when the ITER Agreement was signed in 2006—are not included in U.S.

ITER Project cost estimates because the U.S. ITER Project only covers U.S. contributions to ITER's construction phase.

[16] See appendix III for additional information on the status of each U.S. hardware component as of August 2013.

[17] The European Union, as the host of the ITER site, is not permitted to withdraw from the project under the ITER Agreement.

[18] Department of Energy, *Program and Project Management for the Acquisition of Capital Assets*, DOE Order 413.3B (Washington, D.C.: Nov. 29, 2010).

[19] In December 2012, the Deputy Secretary of Energy authorized the removal of the U.S. ITER Project's designation as a "capital asset project" as defined by DOE's project management order because he determined that the U.S. ITER Project did not fit the definition of such a project. However, the Deputy Secretary directed the Office of Science to continue to apply DOE's project management policies, principles, and processes to the extent practical. According to DOE documents, DOE is following the process described in DOE's project management order, including its major milestones.

[20] For the purposes of this report, we refer to the definitive cost and schedule estimates that DOE sets when it approves a project's performance baseline as "final" estimates. The estimates set at the time DOE approves a performance baseline can change, but DOE's project management order states that such a change would represent an irregular event that should be avoided to the maximum extent possible.

[21] Department of Energy, *U.S. Department of Energy Performance Baseline Guide*, DOE Guide 413.3-5A (Washington, D.C.: Sept. 23, 2011).

[22] According to DOE documents, the U.S. ITER Project will be considered to be complete when DOE has made the last delivery of hardware components and the final cash contribution for the U.S. share of ITER's construction budget. DOE will still be responsible for making cash contributions to the ITER Organization for the operation, deactivation, and decommissioning phases of the ITER project.

[23] Under DOE's program management order, projects are organized by phases and major milestones, starting with the initiation phase and approval of broadly-stated mission needs, known as "CD-0." In 2005, DOE estimated that the U.S. ITER Project would cost $1.122 billion as part of the project's CD-0 milestone. The $1.122 billion cost estimate was a rough-order-of-magnitude estimate, which means it was preliminary and based on high-level project objectives. It remained DOE's cost estimate for the U.S. ITER Project when the ITER Agreement was signed in 2006.

[24] In 2008, DOE approved the start of the U.S. ITER Project's execution phase at the project's second major milestone, known as "CD-1." This milestone included DOE's approval of a preliminary cost range for the U.S. ITER Project of from $1.45 billion to $2.2 billion.

[25] According to DOE officials, the U.S. ITER Project estimate was capped at $2.2 billion until April 2011. They explained that, from January 2011 through August 2013, DOE directed the U.S. ITER Project Office to include various funding profiles in its detailed planning.

[26] As of March 2014, the official ITER Organization schedule anticipates ITER achieving first plasma in 2020, but DOE officials told us this date is unrealistic. Instead, the U.S. ITER Project Office developed the August 2013 estimates based on its best estimate that ITER will achieve first plasma in 2023, with contingency for another year delay to 2024. DOE officials told us that the August 2013 estimates also assumed an increase in the ITER Organization construction budget of 1 billion euros, and the U.S. share of this value was included in the estimates.

[27] The percentages in this list do not sum to 100 percent because of rounding.

[28] According to DOE, the current cost estimate includes 47 percent in contingency for the remaining work to procure and deliver U.S. hardware components, and 12 percent in contingency for U.S. cash contributions to the ITER Organization.

[29] The current $3.915 billion estimate does not include $667 million that was included in the $2.2 billion upper range estimate in 2008 to cover potential future cost growth from risks that

were not fully understood at that time or difficult to quantify, according to DOE documents and officials.

[30] See appendix III for the status of U.S. hardware components as of August 2013. Overall, U.S. hardware component designs were about 52 percent complete as of February 2014, with the level of completion varying by hardware component from as low as 40 percent complete to as high as 100 percent complete. At that time, DOE officials expected to finish delivering the first of the 12 U.S. hardware components in 2015 and to finish delivering the last of the U.S. hardware components in 2030.

[31] DOE officials told us that ITER's design was less developed than the United States understood it to be when the ITER Agreement was signed in 2006. At that time, the United States understood that ITER's design was about 80 to 90 percent complete. However, DOE officials said it was later determined that ITER's design was only about 40 percent complete and that extensive changes would be needed to the design.

[32] State had a role in the negotiation of the ITER Agreement and has a continuing role interpreting U.S. obligations under the agreement. OMB, as the implementation and enforcement arm of presidential policy, is responsible for, among other things, the development and execution of the federal budget. Under the National Science and Technology Policy, Organization, and Priorities Act of 1976 (Pub. L. No. 94-282), the primary function of the OSTP Director is to provide, within the Executive Office of the President, advice on the scientific, engineering, and technological aspects of issues that require attention at the highest level of government. Further, the office serves as a source of scientific and technological analysis and judgment for the President with respect to major policies, plans, and programs of the federal government.

[33] The additional hardware responsibilities DOE agreed to take on were the disruption mitigation systems and the design, fabrication, and testing of insert coils for the central solenoid magnet system and the toroidal field conductor.

[34] GAO-09-3SP and GAO-12-120G.

[35] GAO-09-3SP.

[36] GAO-12-120G.

[37] A sensitivity analysis describes how much costs can change by varying major assumptions, parameters, and data inputs.

[38] We chose to analyze the schedule for the central solenoid modules, which are the largest part of the central solenoid magnet system that will help control the shape of the ITER plasma, and the tokamak cooling water system, which is designed to cool ITER's systems, because they are the largest two hardware items, in terms of value, that the United States is responsible for contributing to ITER. The U.S. ITER Project Office developed the current estimates for these schedules in August 2013.

[39] Horizontal traceability verifies that activities are arranged in the right order for achieving aggregated products or outcomes. Vertical traceability means that varying levels of activities and supporting subactivities can be traced to each other enabling different groups to work to the same master schedule.

[40] DOE officials told us the U.S. ITER Project Office used its best estimate of a realistic international project schedule to develop the August 2013 cost and schedule estimates for the U.S. ITER Project and included a contingency for 1 year of additional potential delay.

[41] At the request of the Office of Fusion Energy Sciences, the Office of Project Assessment within DOE's Office of Science conducted an internal peer review of the U.S. ITER Project in August 2013. The review evaluated, among other things, the project's cost and schedule estimates and the supporting documentation.

[42] DOE, in its fiscal year 2015 budget request, identified the internal peer review's suggested cost range of $4 billion to $6.5 billion as the best estimate for the U.S. ITER Project's potential cost. According to DOE officials, the internal peer review's cost range was not a detailed cost estimate like the U.S. ITER Project Office's $3.915 billion estimate. Rather, the range reflected amounts added to the $3.915 billion estimate by peer review officials to address

their assessment that the U.S. ITER Project Office estimate did not sufficiently consider all project risks and uncertainties. Four of the 10 fusion energy and project management experts we interviewed said they thought the internal peer review's cost range was overstated and unrealistic, with 2 of the experts stating that the review committee used a worst case scenario, which they do not view as likely, to develop the higher end of the cost range.

[43] DOE officials told us that the interim 2-year plan also allows the agency to formally monitor project progress.

[44] The ITER Council Management Advisory Committee, at the request of the ITER Council, reviews and evaluates matters of strategic management importance to the ITER project and makes recommendations to the ITER Council to facilitate the successful development of the ITER Organization and the successful execution of the ITER construction project.

[45] Robert Iotti, a U.S. representative, became the ITER Council Chair in January 2014 and will serve a term of up to 4 years.

[46] This includes amounts identified by DOE as "cost avoidance" or "cost savings." According to OMB, "cost avoidance" is an action taken in the immediate time frame that will decrease costs in the future, and "cost savings" are a reduction in actual expenditures below the projected level of costs to achieve a specific objective. For purposes of this report, we are using the term "cost savings" to encompass both types of actions.

[47] According to DOE's project management order, value engineering is a structured technique commonly used in project management to optimize the overall value of the project. Often, creative strategies will be employed in an attempt to achieve the lowest life-cycle cost available for the project. Value engineering is a planned, detailed review and evaluation of a project to identify alternative approaches to providing the needed assets.

[48] A DOE official told us that, in the absence of a completed strategic plan, the agency believes it has clearly stated the values that are driving U.S. fusion program priorities through DOE's annual budget requests and presentations to stakeholders. Those presentations have provided some information on the program's direction and the challenges it faces, but they are broad in nature and are not equivalent to a completed strategic plan.

[49] See Explanatory Statement, 160 Cong. Rec. H 878 (daily ed., Jan.15, 2014), to the Energy and Water Development and Related Agencies Appropriations Act, 2014, contained in Division D of the Consolidated Appropriations Act, 2014, Pub. L. No. 113-76; and H.R. Rep. No. 112-331, at 855 (Dec.15, 2011) (Conf. Rep.). The fiscal year 2014 explanatory statement further directed that the strategic plan DOE submitted should assume U.S. participation in ITER and assess its priorities for the domestic fusion program based on three funding scenarios.

[50] The Fusion Energy Sciences Advisory Committee provides independent advice to the Director of DOE's Office of Science on complex scientific and technological issues that arise in the planning, implementation, and management of the fusion energy sciences program. Committee members are drawn from universities, national laboratories, and industrial companies.

[51] For example, see GAO-13-115 and GAO-12-77.

[52] GAO-09-3SP and GAO-12-120G.

[53] GAO designed the *Cost Guide* to be used by federal agencies to assist them in developing reliable cost estimates and also as an evaluation tool for existing cost estimates. To develop the *Cost Guide*, GAO cost experts assessed measures applied by cost-estimating organizations throughout the federal government and industry and considered best-practices for the development of reliable cost-estimates.

[54] Not met– DOE provided no evidence that satisfies any of the criterion; minimally met– DOE provided evidence that satisfies a small portion of the criterion; partially met– DOE provided evidence that satisfies about half of the criterion; substantially met– DOE provided evidence that satisfies a large portion of the criterion; and fully met– DOE provided complete evidence that satisfies the entire criterion.

[55] The *Schedule Guide* is intended to expand on the scheduling concepts introduced in the *Cost Guide* by providing 10 best practices to help managers and auditors ensure that the program schedule is reliable. The reliability of the schedule determines the credibility of the program's forecasted dates for decision making.

[56] The Fusion Energy Sciences Advisory Committee provides independent advice to the Director of DOE's Office of Science on complex scientific and technological issues that arise in the planning, implementation, and management of the fusion energy sciences program. Committee members are drawn from universities, national laboratories, and industrial companies.

[57] For example, see GAO-13-115 and GAO-12-77.

[58] The ITER Council Management Advisory Committee, at the request of the ITER Council, reviews and evaluates matters of strategic management importance to the ITER project and makes recommendations to the ITER Council to facilitate the successful development of the ITER Organization and the successful execution of the ITER construction project.

[59] The ITER Council Science and Technology Advisory Committee, at the request of the ITER Council, reviews and advises the ITER Council on science and technology issues concerning the implementation of the ITER project.

[60] We initially selected 15 experts using this multistep process. We eliminated 3 of the initially selected experts because they were working directly on the U.S. ITER Project, either for DOE or the U.S. ITER Project Office. Another expert we initially selected told us she did not have sufficient knowledge of the U.S. ITER Project to participate in our review, and we were unable to get in contact with one other expert we initially selected, so we eliminated both experts and ended up with our final list of 10 experts. We determined that the remaining 10 experts still ensured sufficient coverage and a range of perspectives.

Index

K

L

M

N

O

P

Q

R

S